D0145189

Non docet inſtabiles Copernicus ætheris orbes,
Sed terræ jnſtabiles arguit ille vices.

NICOLAUS COPERNICUS

(From P. Gassendi, *N. Copernici vita*, 1654)

COPERNICUS
and Modern Astronomy

ANGUS ARMITAGE

DOVER PUBLICATIONS, INC.
Mineola, New York

Bibliographical Note

This Dover edition, first published in 2004, is an unabridged republication of the work originally published by Thomas Yoseloff, New York, in 1957 under the title *Copernicus: The Founder of Modern Astronomy.*

Library of Congress Cataloging-in-Publication Data

Armitage, A. (Angus), 1902–
    Copernicus and modern astronomy / Angus Armitage.
      p. cm.
    Previously published: Copernicus, the founder of modern astronomy. New York : Thomas Yoseloff, 1957.
    Includes bibliographical references and index.
    ISBN 0-486-43907-0 (pbk.)
    1. Copernicus, Nicolaus, 1473–1543. 2. Astronomy–History. I. Armitage, A. (Angus), 1902– Copernicus, the founder of modern astronomy. II. Title.

QB36 .C8A7 2004
520'.92–dc22
[B]

                    2004056022

Manufactured in the United States of America
Dover Publications, Inc., 31 East 2nd Street, Mineola, N.Y. 11501

Preface

THIS VOLUME IS THE OUTGROWTH OF A BOOK ON THE SAME SUBJECT published many years ago in England. It was hoped that such an account of the astronomer Copernicus and of the epoch-making work in which he laid the foundations of the heliocentric planetary theory would appeal not only to students of the history of astronomy but also to a wider class of readers interested in Copernicus as one of the makers of modern thought. I have taken the opportunity afforded by the preparation of this new study not only to revise the text of the earlier book and to expand it at several points but also to include two additional chapters.

By way of introduction to the work of Copernicus, it has been necessary to summarize the development of planetary theories from Babylonian times down to his day; this is done in Chapter I. Chapter II contains a concise biography of Copernicus, with some account of his instruments and of the results of the critical examination of his manuscripts. Chapters III to VI consist of an exposition of the contents of Copernicus' book *De revolutionibus*. This work has been studied comparatively with Ptolemy's *Almagest;* and numerous references to the latter have been inserted in order to bring out the close connection between the two astronomical classics. Chapter VII, following the fortunes of the Copernican theory until its establishment by Newton, now takes the place of the brief Epilogue in the earlier work. An eighth chapter has been added dealing with the historic attempts to verify physically the diurnal and orbital motions of the earth. The Bibliography has also been revised.

When I first embarked upon the study of Copernicus, I received much valuable help from my teachers, the late Professors A. Wolf and L. N. G. Filon, and from the late Professor H. E. Butler, all of University College, London; more recently I have benefited from the criticisms and from the published researches of Dr. Edward Rosen, of the College of the City of New York. To these, and to many other authorities, I acknowledge my debt; but the responsibility for any errors or omissions in the following pages must remain solely mine.

A. A.

Note

IN THE PRESENT WORK NUMEROUS REFERENCES WILL BE MADE TO various passages in Copernicus' book *De revolutionibus orbium coelestium* and in Ptolemy's *Almagest*. Such references to a given book and chapter of the *De revolutionibus* will, in general, simply be indicated thus: (III, 15); references to a given book and chapter of the *Almagest,* thus: (*Alm.,* XI, 7); and cross references to a given chapter and section of this book, thus: (Chapter I, § 2, *supra*) or (Chapter IV, § 3, *infra*).

Contents

Plates

If they list to try
Conjecture, he his Fabric of the Heavens
Hath left to their disputes—perhaps to move
His laughter at their quaint opinions wide
Hereafter, when they come to model Heaven,
And calculate the stars; how they will wield
The mighty frame; how build, unbuild, contrive
To save appearances; how gird the Sphere
With Centric and Eccentric scribbled o'er,
Cycle and Epicycle, Orb in Orb.

MILTON, *Paradise Lost,* Book VIII

COPERNICUS
and Modern Astronomy

1

Planetary Theories Before Copernicus

THE RECORDED HISTORY OF ASTRONOMY UNTIL THE SEVENTEENTH century is occupied not so much with discoveries of previously unknown phenomena as with a succession of attempts to systematize and to interpret certain facts of which the earliest recognition lies altogether beyond the horizon of history. The ancient and medieval schools of astronomy were concerned with celestial processes which had been conspicuous to unaided human perception from of old. For ages the stars had been observed to form permanent groups, or constellations, that wheeled in hourly and seasonal sequences across the vault of the night sky. The moon had been watched, waxing and waning on its monthly circuit through the central belt of the constellations, and the sun, varying its daily course, and rising and setting with different stars, according to the season of the year. Less conspicuous than sun or moon, but nevertheless distinguished from the stars since prehistoric times, were the planets, circulating slowly and erratically through the constellations. Add to these such occasional spectacles as eclipses and the visitations of comets, and we have before us all the principal phenomena with which astronomers were concerned prior to the invention of the telescope some three hundred and fifty years ago.

To the age-long quest for system and significance in these celestial phenomena a fresh turn was given in the sixteenth century by the astronomer Nicolaus Copernicus, whose contribution to solving the classic problem of the cosmos forms the central theme of this book. Copernicus inherited from antiquity a fully developed philosophy of nature which, on the one hand, conditioned his own novel approach to the problem, and, on the other, served as the basis for the long-continued opposition to his views. To his predecessors also he owed both the geometrical technique which he employed to represent the planetary motions and many of the recorded observations which he utilized to define the constants of his geometrical theory. Unless we know something of the history of astronomy in the ages before Copernicus, we shall scarcely follow his arguments and calculations or judge his achievements aright. Accordingly, we shall devote this opening chapter to a brief preparatory study of such developments in ancient and medieval cosmology as are relevant to what follows.

§ 1. ANTIQUITY

The mighty stream of modern astronomy can be traced back to the confluence of two tributaries derived from the contrasting civilizations of Babylonia and Greece. Even at the level of barbarism, practical life must have demanded of man some knowledge of the basic celestial phenomena, for direction-finding and for regulating the agricultural and ritual cycle of the year. However, conditions favorable to the germination of science are first found in the historic civilizations which arose on the alluvial lands of the great rivers of antiquity, notably in Mesopotamia.

The Babylonian priests at first observed the heavens in order to regulate their lunar calendar and to draw omens from all striking celestial or atmospheric phenomena. The stepped towers, or *ziggurats,* of the temples were their observatories; and through the stormy centuries of Babylonian history the temple schools were able to develop elaborate systems of star lore, since invaders generally spared the national shrines of the old gods of the land. During the past century a growing light has been shed upon the astronomical achievements of the Babylonians by the decipherment of the clay tablets upon which their observations, calculations, and ephemerides were recorded. In the earliest stage, observations were, for the most part, indiscriminate and lacked numerical precision, the phenomena being treated merely as portents of impending crises. The next stage, well established by the eighth century B.C., was characterized by dated records, numerical specifications, and estimates of the periods of recurrent celestial phenomena. The discovery of periodicity marked the beginning of scientific astronomy; it led, in the three centuries preceding the Christian era, to the highest level of Babylonian astronomical attainment, represented by the construction of ephemerides that served to predict celestial phenomena years in advance.

The Babylonians attached especial significance to the movements of the seven "planets," namely, the sun, the moon, and the bodies to which (following the Romans) we give the names of Mercury, Venus, Mars, Jupiter, and Saturn, and which we class as planets in the modern sense of the term. They associated or identified their chief gods with these bodies. The planetary phenomena to which they paid particular attention were (a) the *heliacal risings and settings* of a planet, when it was observed to rise before sunrise for the first time or to set after sunset for the last time (occasions which marked the limits of its period of extinction in the sun's rays); (b) the *stationary points* where a planet's course among the constellations was arrested and reversed; (c) *oppositions*, when the planet was in the opposite quarter of the sky to the sun; (d) *conjunctions*, when the planets appeared to pass close to one another or to bright stars; and (e) eclipses of the sun or moon.

The phenomenon of *stationary points* calls for some further explanation, because of the complications which it necessitated in planetary theories until the time of Copernicus. The sun and moon, in their periodic circuits round the heavens, travel con-

tinuously through the constellations from west by south to east;
but the apparent motions of the five planets are more complicated. For example, if the planet Mars is observed in the southern sky night after night for some weeks, it will, in general, be
found to be slowly moving from west to east in relation to the
background of stars. (Such motion is, of course, to be distinguished from the apparent diurnal revolution about the earth,
which the planet has in common with the stars.) At fairly regular intervals (about 780 days for Mars), however, the planet's
eastward motion is arrested (its apparent path showing a *stationary point*) and reversed, and Mars moves from east to west
through an *arc of retrogression* of about 15° before resuming its
normal eastward motion. The same is true of the planets Jupiter and Saturn, although their arcs of retrogression are less considerable; the planets Mercury and Venus have the additional
peculiarity that the angular distance of each from the sun
never exceeds a moderate limiting value (about 25° for Mercury and 45° for Venus).

From at least the beginning of the fourth century B.C. the
Babylonians were in the habit of defining the apparent positions
of the planets in the sky by reference to a series of bright stars
distributed fairly regularly round that belt of the heavens in
which all the planets move, and which (following the Greeks)
we term the *zodiac*. The place of a planet was defined by the
specification of its angular separation from the standard star
nearest to it. A scale of angular measurement was afforded by
the division of the zodiac into twelve equal parts (the *signs* of
the zodiac), with further subdivisions, and the system was gradually developed so that such measurements might be made both
along the zodiac and at right angles to it. This system led eventually to the conception of the *ecliptic*—the great circle of the
celestial sphere traversed by the sun in its annual circuit through
the constellations. Thus, by the second century B.C., the *celestial
longitudes and latitudes* of the planets and of other celestial

bodies were being defined by reference to the circle of the ecliptic, much as we define the geographical longitudes and latitudes of places on the earth by reference to the terrestrial equator.

By diligent observations, continued over several centuries, the Babylonians were able to arrive at remarkably accurate estimates of the principal *time periods* associated with the heavenly bodies—the year, the several kinds of months, the *sidereal period* of each planet (in which, on the average, the planet performs a complete circuit of the heavens relative to the background of stars), and its *synodic period* (the time for a complete circuit relative to the sun). By the second century B.C., the Babylonians had arrived at an estimate of the sidereal year: 365 days, 6 hours, 13 minutes, 43.4 seconds, only about four and a half minutes in excess of the modern estimate for that age (F. X. Kugler: *Sternkunde und Sterndienst in Babel,* II, 8). And about the fourth century B.C. they had already discovered that lunar eclipses form sequences which recur periodically at intervals of about eighteen years. In their intricate ephemerides of the sun, moon, and planets, the Babylonians made allowance for the principal periodic non-uniformities observable in the apparent motions of these bodies through the constellations.

It is difficult to give in a few words any idea of the refinement and complexity of the Babylonian methods. Tables for showing the dates of successive new moons (second century B.C.) included columns of corrections for the yearly inequality in the sun's motion and for the monthly inequality in the moon's motion; the amounts of the corrections for successive months oscillated in a regular manner between maximum and minimum values. As regards the remaining five planets, the Babylonians took account not only of the element of inequality in the motion of each which depends upon its position in relation to the sun (and which we now know to be due to the earth's motion) but also of the further element, complicating the first, which depends upon the planet's position in the zodiac (and which results from the planet's elliptic motion). They represented this latter inequality by assigning to the planet, as it moved round the zodiac, a succession of rates of angular motion, which varied between maximum and minimum values and recurred in the appropriate period.

Thus, in contrast to the geometrical and physical astronomy with which we shall be concerned in the following pages, Babylonian planetary theory, in its classic form, is represented merely by tables in which the motions, past and future, of the heavenly bodies are numerically formalized in complete abstraction from any physical or even geometrical conceptions of the cosmos.

It is unlikely, in fact, that the Babylonians ever arrived at any clear-cut, objective conception of the constitution of the universe as a whole, such as we encounter in Greek philosophy. It is possible, however, to piece together a composite picture of the cosmology which must have formed the background of Babylonian thought, at least throughout the historic period. The earth was conceived to be roughly circular in contour, rising toward the center to form a huge mountain, and resting upon a great ocean which girdled the land with a moat of sea; beyond this rose a circular mountain wall, forming the boundary of the world and supporting the hemispherical vault of heaven, or firmament. The heavenly bodies seem to have been regarded as moving freely through space.

The earliest extant Greek literature—the poems of Homer and Hesiod—presupposes a system of the world closely resembling that of the Babylonians. But Greek cosmology soon came to reflect the sharp contrast existing between Babylonian civilization and Greek society prior to the Alexandrian period. Even in their homeland the Greeks were never subject to a conservative and centralized priesthood; and in any case population pressure drove many of them to found colonies overseas, particularly in Asia Minor and in southern Italy. In the Ionian trading cities of the Asian coast there flourished, during the sixth and fifth centuries before Christ, the earliest known schools of philosophy. These schools established the conception of nature as an orderly system the constitution and phenomena of which were not to be attributed to supernatural agencies but were to be rationally deduced as consequences of the inherent properties of the one or more primary substances of which the entire universe was held to be composed. The Ionian philosophers, in fact, interpreted cosmic phenomena on the analogy of the natural processes with which they were familiar through the technical arts of their own largely industrial society. They elaborated a succession of naturalistic, if somewhat crude and speculative, cosmological systems, in which the earth generally figured as a disc, or a vortical condensation, floating at the center of the universe. The conception, attributed to Anaximenes of Miletus, of a rotating crystal sphere to which the stars are attached like silver studs persisted from the sixth century B.C. to the end of the sixteenth century A.D.

Meanwhile, at another outpost of Greek civilization in southern Italy, Pythagoras and his school were establishing a scheme of geometrical abstractions in terms of which celestial processes would continue to be conceived for upwards of two thousand years. To Pythagoras himself (sixth century B.C.) is credibly attributed the earliest declaration that the earth is a sphere resting at the center of a spherical universe, and not a floating disc, as the Ionians taught. Pythagoras seems also to have been aware (probably following the Babylonians in this) that the complicated apparent motion of the sun in the course of the year is, to all appearance, made up of two simple motions—(a) the motion common to all the heavenly bodies, whereby they appear to revolve about an axis through the earth once in a day, and (b) a motion peculiar to the sun, in a contrary direction to the first motion, and taking place about a different axis, in the space of one year. The first of these motions would account for the daily rising and setting of the sun; the second would account for the sun's annual circuit among the constellations and for the seasonal fluctuations in its rising and setting points and in its meridian altitude. This analysis seems also to have been extended by Pythagoras (although with less eligibility) to the apparent motions of the moon and planets; it gave rise to the idea that the complicated movements of the heavenly bodies could all be resolved into uniform circular motions. This doctrine was established by the authority of Plato, Aristotle, and Ptolemy; as we shall see, it still dominated astronomy in the time of Copernicus, two thousand years after Pythagoras, and it was first formally abandoned by Kepler at the beginning of the seventeenth century.

By the end of the fifth century B.C., the Pythagorean school had evolved the remarkable system of cosmology associated with the name of Philolaus. To this hypothesis Copernicus directed especial attention, for it was the earliest historic system to displace the earth from the center of the universe and to set it in revolution about the center like any other planet. According to Philolaus, the finite sphere within which the universe was contained had fire at its center and fire at its circumference. It was divided by concentric spheres into three layers. The outermost of these contained the stars. The intermediate layer contained the five planets, the sun, and the moon, in order of approach to the Central Fire, about which these bodies all revolved in circles in the several periods of their circuits round the heavens. Lastly, within the sphere which formed the core of the universe was the earth, which itself revolved daily about the Central Fire, turning toward it the hemisphere opposite to that inhabited by mankind, who, in consequence, could never behold the Fire. By making the earth revolve in a plane inclined to that in which the other planets moved, Philolaus was able to account not only for the risings and settings of the heavenly bodies but also for all seasonal phenomena now attributed to the inclination of the earth's equator to the ecliptic.

The somewhat fanciful system of Philolaus never established itself; but the primitive Pythagorean cosmology lived on in the natural philosophy of Plato (427–347 B.C.), who conceived the universe as a rotating sphere in the midst of a boundless void, and the earth as a stationary sphere in the midst of the universe. In picturesque allegories in the *Timaeus* and the *Republic,* it is easy to recognize Plato's expressions of the principle that the motion of each of the seven planets is compounded of two uniform revolutions about the earth which take place about different axes and in opposite senses, and one of which is common to all the heavenly bodies. It was obvious, however, that this simple Pythagorean conception of a planet's motion took no account of the recurrent *retrogressions* of the planets, to which we have already alluded, or of their departures from the ecliptic. It was probably with a view to remedying this defect in the current theory that, according to Simplicius, Plato set the astronomers of his time the general problem of adequately representing the observed movements of the heavenly bodies by combinations of uniform circular motions having a common center in the earth.

This problem was taken up, early in the fourth century B.C., by Eudoxus of Cnidos, a pupil of Plato and one of the greatest mathematicians of antiquity. Eudoxus regarded each planet as attached to the equator of an ideal sphere, which rotated uniformly about two opposite poles, with the earth at its center. The poles of this sphere were embedded in the surface of a second sphere, external to the first, but concentric with it, and itself in uniform rotation about an axis inclined at a constant angle to that of the first. This second sphere was similarly related to a third one, and so on. Eudoxus' problem was, then, to choose, for each planet, a particular combination of these spheres, having such axes and periods of rotation that the super-

position of their motions would make a point on the equator of
the innermost sphere move about the common center with a
motion similar to that with which the planet in question was
observed to move about the earth. The system of each planet
had one sphere the function of which was to impart the diurnal
motion about the polar axis of the heavens common to all heavenly bodies. In the system of the moon, there was a second
sphere the rotation of which corresponded to the monthly eastward revolution of the moon in the plane of its orbit, and a
third sphere which provided for the slow westward regression
of the *line of nodes* in which that orbit intersects the ecliptic.
(That the moon's path was inclined to the ecliptic, intersecting
it at two *nodes*, was proved by the fact that eclipses do not take
place at every new and full moon; and these nodes, near which
all eclipses occur, were found to revolve round the ecliptic from
east to west.) An analogous group of three spheres was postulated for the sun, which was mistakenly believed to deviate from
a great circle in its annual course through the heavens. In the
application of his theory to the motions of the five planets, Eudoxus was able to give a fair representation of the characteristic
motion of a planet among the constellations. In addition to two
spheres respectively conferring upon the planet its diurnal revolution and a uniform eastward revolution in its own sidereal
period (as in the simple Pythagorean theory) Eudoxus introduced two further spheres—the first of these rotating about an
axis lying in the plane of the ecliptic and the second, about an
axis inclined at a constant angle to that of the first, in the same
period (the synodic period of the planet, see p. 21), but in the
opposite sense of rotation. The effect of imparting these two
motions to the planet would be to make it describe about its
mean position a curve like a figure eight lying on its side. This
motion, when superimposed upon the uniform eastward motion
already possessed by the planet, would give rise to such periodic
retrogressions as are actually observed, provided the inclination

of the axes of the two last-mentioned spheres were suitably chosen. The departures of the planets from the plane of the ecliptic involved in Eudoxus' hypothesis, however, had no relation to the motions in latitude which are actually observed; and the hypothesis was not applicable, in its original form, to the planet Mars, whose arcs of retrogression are more considerable than those of the more distant planets. Nor did the system of Eudoxus take account of the effects now known to arise from the eccentricity of planetary orbits, which, in the case of the sun's apparent motion, produces the inequality of the four seasons.

Some of the defects in Eudoxus' theory were removed by his immediate successors; but one fundamental objection against the whole theory remained: it took no account of variations in the distances of the heavenly bodies from the earth, such as are suggested by variations in the brightnesses of the planets and by the occurrence of total and of annular solar eclipses, which proves that the relative distances from us of the sun and moon are liable to vary. The spheres of Eudoxus, however, determined the pattern of Aristotle's cosmological system, the main outlines and regulative physical principles of which retained, down to the age of Copernicus, an authority altogether disproportionate to their value.

Aristotle (384–322 B.C.) conceived the physical universe as a finite sphere embracing all that exists. At its center was the earth, a stationary sphere, round which the rest of the universe was built up symmetrically in concentric spherical shells. About the central mass of earth lay successive layers occupied predominantly by water, air, and fire, respectively, which completely filled the region comprised within the sphere carrying the moon. The remainder of the universe, to the outermost sphere carrying the fixed stars, was occupied by the successive systems of planetary spheres, which Aristotle regarded as physically real and in mutual contact, each sphere transmitting its motion to

the one next within it. The moon's sphere was supposed to separate two fundamentally different regions of the universe.
Within this sphere all things were composed of the four elements, earth, water, air, and fire, which were constantly undergoing transformation one into another, in virtue of their common substratum of formless primary matter, so that the sub-
lunary region was characterized by incessant generation, change,
and decay. Beyond the moon's sphere, however, the celestial
bodies and their carrying-spheres were all composed of an incorruptible fifth element, or *ether,* capable of undergoing only
change of *place.* This doctrine of the immutability of the superlunary realm was still current in the sixteenth century. It was
reluctantly abandoned in the seventeenth century, when comets
and "new stars" had been definitely admitted to be celestial
phenomena, whereas the distinction between terrestrial and
celestial matter was conclusively abolished only after the development of spectrum analysis, in the nineteenth century. Aristotle's dichotomy of elemental and celestial bodies involved a
fundamental difference in their natural modes of motion. It
was upon his theory of motion that Aristotle based his arguments for retaining the earth as a stationary mass at the center
of the universe; but we shall deal with these arguments later,
in connection with Copernicus' attempts to refute them (see
Chapter III, § 3, *infra*).

Among the pupils of Plato and Aristotle in the fourth century
B.C., Heraclides of Pontus must be mentioned here. For he
taught that the apparent diurnal revolution of the heavens was
actually due to an axial rotation of the earth in the opposite
direction. This hypothesis is also attributed by certain classical
writers to Ecphantus and Hicetas, both Pythagoreans of Syracuse (dates unknown, but probably anterior to Aristotle), to
whom Copernicus alludes in his book. The relation of these
men to Heraclides is obscure. It is suggested by Heath that
Ecphantus may have been introduced by Heraclides as a typi-

cal Pythagorean into one of his lost dialogues, to serve as a mouthpiece for the writer's own opinions. The ascription of the hypothesis to Hicetas may be due to a similar confusion.

Heraclides also correctly explained the curious property of Mercury and Venus whereby each appears alternately east and west of the sun and never recedes far from it. He suggested that these planets revolved in circles about the sun as center, while the sun revolved in a larger circle about the earth. It has been suggested that this limited heliocentric system (wrongly attributed by the Latin commentator Macrobius to the Egyptians, and hence sometimes known as the "Egyptian system") was probably extended (by whom is unknown), during the century following Heraclides, to include Mars, Jupiter, and Saturn, so that all the five planets would thus be supposed to revolve about the sun while the sun revolved about the earth. (In this system the superior planets would be nearest to the earth when in *opposition* to the sun; this would explain why, in fact, they appear brightest in that situation.) This would have marked an important step toward the Copernican scheme; and, early in the third century B.C., Aristarchus of Samos actually anticipated the full Copernican system in its broad outlines. That is to say, he put forward the hypotheses that the sphere of stars was motionless, so that its apparent daily revolution was due to a diurnal rotation of the earth; that the sun was at rest at the center of the sphere of fixed stars; that the earth and planets described circles about the sun as center; and that the radius of the sphere of stars was so incomparably greater than that of the earth's orbit that no corresponding apparent motion was produced in the stars. The hypotheses of Aristarchus, however, were generally rejected as impious and contrary to sound physical principles. Moreover, such undeveloped speculations had soon to compete with the carefully articulated systems of the Alexandrian astronomers, which represented inequalities in the apparent motions of the sun and planets with steadily increasing numerical accuracy. These systems were conceived on the geocentric hypothesis, and their success did much to establish it.

The development of orthodox planetary theories for a century or so after Heraclides is obscure. We have already mentioned the possibility that his hypothesis was extended from the inferior to the superior planets during the third century B.C. In such a system, the distinction between these two classes of planets would be that the circle described about the sun by an inferior planet would be smaller than the circle described by the sun about the earth, whereas that described about the sun by a superior planet would be larger than the sun's circle and would embrace the earth. It was probably by generalizing these two alternative kinds of motion of a planet about the earth that the Greeks arrived at two geometrical devices for representing such motion which were to be extensively employed in all planetary theories down to the time of Kepler. As we shall repeatedly encounter these devices in our study of Copernicus, we must now explain their nature (see Fig. 1; no account is taken, in this explanation, of the diurnal motion common to all the heavenly bodies).

For the sake of completeness, we begin with the ideally simplified planetary system of Pythagoras and Plato, in which a planet, P, described a circle at a uniform rate about the earth, E, at the center (Fig. 1, i). This system was developed, as we have seen, into the spheres of Eudoxus, which were merely combinations of such geocentric circular motions, and in which no account was taken of variations in the distances of the planets from the earth. From the third century B.C., however, such variations in the distance, as well as in the rate of apparent motion, of a planet, were represented in either of two ways. In the first of these (Fig. 1, ii) the planet P was conceived as uniformly describing a circle (called a *movable eccentric*) about a center, C, which itself meanwhile described a circle about the earth, E. This system was most obviously suited to represent the motions

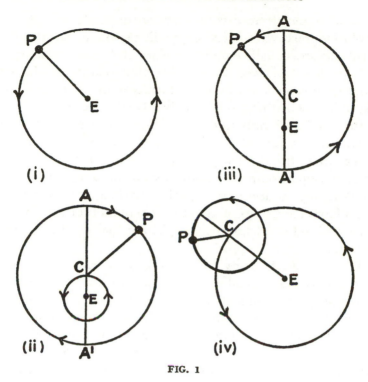

FIG. 1

of the superior planets, the point C (which might be identified with the sun) revolving about the earth from west to east in one year, relative to the stars. The planet P performed a complete circuit round the earth from west to east (relative to the stars) in its sidereal period. But the Greeks regarded the planet's circle, APA', as being carried round E by the motion of C, as if it were a material circle attached to the moving line ACA'; and they reckoned the motion of the planet on the circle not in relation to the stars but from the moving point A (*apogee*) where it was farthest from the earth. Hence, in order just to complete a circuit about the earth from west to east in its *sidereal* period, the planet had to be supposed to travel in a retrograde direction (from east to west) on its circle, completing a revolution, relative to A, in its *synodic* period. A particular case of (ii) was occasionally employed in which C was stationary and the line AA′ preserved a fixed direction relative to the stars, while the planet P revolved uniformly about C from west to east in its sidereal period (Fig 1, iii). Lastly, the planet P could be supposed uniformly to describe a small circle (called an *epicycle*) about a center, C, which itself meanwhile described a larger circle (the *deferent*) about the earth, E (Fig. 1, iv). By suitably proportioning the radii of such combinations of circles and assigning the appropriate periods to them, it was possible to give a fairly correct geometrical representation of the retrogressions and other characteristic phenomena of the planetary motions.

According to Ptolemy (*Alm.*, XII, 1), Apollonius of Perga (*fl. c.* 230 B.C.), the "great geometer" of antiquity, was acquainted with the systems both of movable eccentrics and of epicycles, and he understood their mathematical properties. Now, according to Heraclides' hypothesis, Mercury and Venus described *epicycles* about the sun (which described a *deferent* about the earth), while, in the supposed extension of this hypothesis, Mars, Jupiter, and Saturn described *movable eccentrics* about the sun, which occupied the point C in Fig. 1, ii (although by the end of the third century B.C. these systems had been generalized, so as to make the planets, and the sun itself, revolve about imaginary points). This gave a heliocentric system, so far as the planets were concerned, and it might soon have been developed into the Copernican system (it probably was the means by which Aristarchus was enabled to take that step), especially as the year figured in each separate planetary theory and was evidently a fundamentally important period. But the planets appeared to fall into two classes, according as they described *eccentrics* or *epicycles;* and it was known to Apollonius that any eccentric system can be transformed into an epicyclic system, and vice versa, by interchanging the radii of the two circles and transforming the respective periods according to definite rules (see Note I, *infra*). Consequently, during the centuries between Apollonius and Ptolemy, the eccentrics of the superior planets gave place to epicycles of which the centers of the deferents coincided with the earth. The reason for this preference was presumably the apparent gain in simplicity involved in the application of the same device (the *epicycle*) to both superior and inferior planets. But the sun and the year no longer figured in all the planetary theories (for, in accordance with the above-mentioned convention, the period of revolution in the epicycle was the *synodic* period of the planet), and the chance of progressing to a heliocentric system was lost for the time being.

The only astronomer known to have made important contributions to the development of planetary theories during the four centuries from Apollonius to Ptolemy was Hipparchus of Rhodes (*fl. c.* 150 B.C.). All the important works of Hipparchus are lost, and his achievements, known to us chiefly through the *Almagest,* cannot always be distinguished with certainty from those of the Babylonians (of whose recorded observations he made considerable use), or even from those of Ptolemy himself.

In order to account for the apparent motion of the sun in the zodiac, with the annually recurring fluctuation in its rate, Hipparchus supposed the sun, S, to revolve on a fixed eccentric circle uniformly about the stationary center, C, which was displaced some distance from the stationary earth, E (Fig. 2; cf. Fig. 1, iii). From a knowledge of the approximate durations of spring and summer and of the length of the year, he was able to calculate the constants of the sun's orbit, viz., the *eccentricity* (the ratio CE : CA) and the direction in relation to the equinoctial and solstitial points, of the *apse line* AA' joining the sun's *perigee,* A, and *apogee,* A'. Ptolemy adopted this representation of the sun's motion, and Copernicus retained it as a

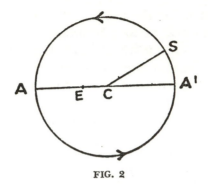

FIG. 2

means of accounting for the principal solar inequality. Hipparchus sought, less successfully, to represent the inequality in the moon's motion by assuming the satellite to describe an epicycle which was carried round upon a deferent intersecting the ecliptic in regressing nodes (cf. p. 26); so far as the five planets were concerned, Hipparchus merely classified and supplemented existing observations of these bodies, thus preparing the way for Ptolemy three centuries later. Hipparchus' reputed discovery of the precession of the equinoxes, his star catalogue, and his attempt to determine the distances of the sun and moon will be referred to later, at the points where they become relevant to our account of Copernicus.

The Babylonians seem to have made allowance (as has already been noted) not only for the inequality in a planet's motion which gives rise to the retrogressions, and which recurs in the planet's *synodic* period, but also for the inequality which depends upon the planet's position in the zodiac, and which recurs in very nearly its *sidereal* period. Hipparchus was apparently aware of this twofold inequality (*Alm.*, IX, 2), and he supposed that, in order to represent it, *combinations* of eccentrics and epicycles would be required. A solution of this problem which survived, in its broad outlines, for fourteen centuries was the essential contribution to planetary theory made by Ptolemy of Alexandria (*fl. c.* A.D. 150).

Ptolemy's mode of representing the motions of Venus and of the three superior planets was, essentially, as follows (Fig. 3):

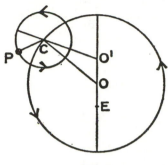

FIG. 3

The planet P described from west to east an epicycle of which the center, C, described in the same sense a deferent of which the stationary center, O, was eccentric to the earth, E. The point C was conceived to move with uniform angular velocity, not about the center O (as in earlier planetary theories), nor even about the earth, E, but about a point, O', lying in EO produced, and such that $EO' = 2 \cdot EO$. The motion attributed to Mercury was considerably more complicated; and there were elaborate schemes for representing the motions of the planets in latitude.

Ptolemy's lunar theory may be summarized as follows (Fig. 4): The moon was supposed to move on an epicycle, MN, the center of which, A, moved from west to east on an eccentric deferent, the center of which, F, in turn, revolved from east to west about the earth, E, the whole lying in the plane, ABCD, of the moon's apparent motion. To an observer at E, the opposite motions of A and F were equal relative to the line AC join-

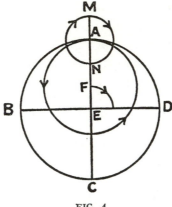

FIG. 4

ing earth and sun. Thus the epicycle was at apogee on the eccentric at the times of new and of full moon, and at perigee at the
time of half-moon. This theory (to which some further refinements were added) made allowance for a periodic fluctuation in
the eccentricity of the moon's orbit, the effects of which seem to
have been first clearly distinguished by Ptolemy, and which is
now known as the *evection* (see Chapter V, § 1, *infra*).

Ptolemy's lunar and planetary theory completed the achievement of Greek and Alexandrian astronomy, the roots of which
lay deep in the ancient civilizations of the East. He wrought the
whole into the comprehensive, logical system of astronomy set
forth in a treatise, the ΜΑΘΗΜΑΤΙΚΗ ΣΥΝΤΑΞΙΣ, now commonly known as the *Almagest*. * The authority of this work
(which was completed about A.D. 145) was solidly on the side of
the geocentric theory, and it dominated all the developments
of astronomy with which we are here concerned until the sixteenth century, when the book served as a quarry from which

* So called after the Arabic title, *Al-majisti*, which was formed either by prefix-
ing the article *al* to the Greek adjective μεγίστη (greatest) applied to the *Syntaxis*,
or, more probably, according to Sarton (*History of Science*, I, 562), from an artificial contraction of the words μεγάλη σύνταξις (great collection), by which the
work was known to the commentators.

Copernicus extracted many of the data and geometrical methods which he employed in his effort to subvert that authority.

The developments in cosmology and planetary theory during the fourteen centuries from Ptolemy to Copernicus were relatively unimportant when compared with those of the seven centuries between Pythagoras and Ptolemy, and they may be dealt with more briefly. A succession of commentators and compilers kept alive the tradition of Alexandrian astronomy for several centuries after the death of Ptolemy. But even such activities ceased almost entirely in Christendom about the beginning of the sixth century, when the Athenian schools of philosophy were closed by the Emperor Justinian.

§ 2. THE MIDDLE AGES

Such developments as took place in cosmology during the Middle Ages were conditioned by the progressive recovery of the submerged classics of ancient science rather than by the slowly reviving practice of celestial observation. The scholars of Western Christendom drew first upon the scanty materials furnished by the Latin Fathers and the Encyclopedists, with their predominantly Platonic tone. Later they encountered the tradition of Islamic learning, which introduced them to the works of Aristotle and Ptolemy in an Arabic setting. Lastly they obtained possession of the original Greek texts and were exposed to the liberating influence of the Humanist movement. These developments will now be reviewed.

The conversion of the Roman Empire to nominal Christianity in the fourth century was followed by the irruption of the barbarians; the tradition of ancient science was broken; and there arose from the ruins of the Western Empire a civilization that derived its ideas about the universe mostly from the teachings of the Church. The earlier ecclesiastical writers had sought to harmonize the cosmology of the Schools with the words of

Scripture (allegorically interpreted where necessary). But, with their rise to power, the leaders of the Church showed a growing intolerance of the doctrines of classical science; from about the fourth to the sixth century, even the spherical form of the earth was frequently denied, and the structure of the universe was sometimes conceived as analogous to the Tabernacle of Moses. In the eighth century, however, the Venerable Bede gave an elementary account of the planetary motions, based upon the particulars in Pliny's *Natural History*. The scientific knowledge available in the West at this period consisted, in fact, largely of what could be gleaned from the writings of the Fathers of the Church and from Pliny's book; it was embodied in a succession of encyclopedias and commentaries. By the end of the ninth century, Christian scholars had become acquainted, through the Latin writings of certain Neo-Platonist commentators, with the cosmological ideas of Plato's *Timaeus* (the influence of which upon medieval Christian thought remained virtually unchallenged until the twelfth century) and with the planetary hypothesis of Heraclides of Pontus (§ 1, *supra*), allusions to which occur throughout medieval literature until the time of Copernicus.

Meanwhile a remarkable revival of the astronomical knowledge and activities of classical antiquity had been occurring in the countries under Islam. On their conquering march westward through Egypt and eastward to the Indus the Arabs came into contact both with what remained of Alexandrian culture and with Hellenic ideas originally implanted in the Middle East through the conquests of Alexander the Great. As the wave of Muslim expansion reached its limits, the Arabs fell under these influences; and scientific pursuits flourished at their courts and academies from the eighth century to about the thirteenth, at Bagdad, in Egypt, and in Moorish Spain. Scholarly Arab potentates had Greek and Hindu scientific works translated into

Arabic (including Ptolemy's *Syntaxis,* about 820); and they founded observatories where important astronomical constants were redetermined, star catalogues were compiled, and tables of planetary motions were computed.

As to how the motions of the planets should be represented, the judgment of Muslim philosophers was divided between the two traditions respectively associated with the names of Aristotle and Ptolemy. The complicated motions attributed by Ptolemy to the planets were clearly inconsistent with the physical principles of planetary motion laid down by Aristotle. Whether the acceptance of Ptolemaic astronomy necessarily involved the rejection of Aristotelian physics, however, depended upon what status was to be assigned to such planetary hypotheses as those in the *Almagest.* Upon this question Muslim philosophers showed a difference of opinion which had already arisen among the late Alexandrian writers and which was to reappear in the schools of Christendom. Some of them considered, with Averroes, that a planetary hypothesis should conform to the physical laws of motion appropriate to celestial bodies; they adhered to the teachings of Aristotle, and they rejected the Ptolemaic system altogether. Among these was the Spanish astronomer al-Bitrugi (twelfth century), who proposed a system of homocentric planetary spheres in which some allowance was made for the precession of the equinoxes; this system proved a serious rival to the Ptolemaic hypothesis and later gained considerable acceptance in northern Europe. Other thinkers regarded planetary hypotheses as mere artifices for systematizing the observed motions of the planets and for computing tables to predict their future movements; they recognized the inadequacy of the conception of homocentric spheres for this purpose, and they accordingly adopted the Ptolemaic system. Still others sought to resolve the conflict by showing that Aristotle's words would bear interpretation broad enough to comprehend all that Ptolemy had postulated concerning the planetary motions.

Several of the Arab disciples of Ptolemy tried to give a physical significance to the details of his system by substituting hypothetical, material mechanisms for the purely geometrical schemes of the *Almagest,* as, indeed, Ptolemy himself had sought to do in one of his later works, Ὑποθέσεις τῶν πλανωμένων (*Planetary Hypotheses*). They supposed each planet to turn upon an epicyclic sphere in the interspace between the concentric spherical surfaces of two ethereal solids which themselves slid freely over other solids. The interspace was eccentric to the earth, and the solids were endowed with such rotations as to provide for the various phenomena of the planets' motions.

One of the greatest of the Arab astronomers, al-Battani, (ninth and tenth centuries), upon redetermining the elements of the sun's apparent orbit, noticed that the longitude of its apogee differed appreciably from that given by Ptolemy; he is therefore generally regarded as the discoverer of the progressive motion of the sun's apses upon the ecliptic. For the rest, despite centuries of concentration upon the details of the Ptolemaic planetary system, the Muslims made no significant improvements upon it in principle. They lacked the mathematical genius and the critical faculty of the Greeks; and, under the influence of Neo-Platonist writings (some of them falsely ascribed to Aristotle), their natural philosophers became involved in fantastic speculations concerning the hierarchy of Intelligences supposed to animate the successive planetary spheres. However, they left many recorded observations that were utilized by succeeding astronomers, including Copernicus. They transmitted to the West our system of numerals and the conceptions of the sine and other trigonometrical functions. Above all, it was the Muslims who kept alive the tradition of ancient science and philosophy and who, during the period of intellectual stagnation in Europe, preserved many of the texts in which that tradition was enshrined.

By the twelfth century this Greco-Arabic science had begun

to infiltrate into Western Christendom, where it produced a considerable widening of the intellectual horizon. Latin translations were made from the Arabic versions of some of the chief scientific and philosophical works of antiquity. These classics included Ptolemy's *Almagest* (translated from the Arabic by Gerard of Cremona in 1175) and the physical and astronomical works of Aristotle. In the thirteenth century the Aristotelian system of natural philosophy established itself in the schools of Western Christendom. It encountered there some initial opposition on theological grounds, since it taught or implied doctrines (such as that of the eternal existence of the universe) that conflicted with Christian, Jewish, and Muslim orthodoxy alike; an ineffective attempt was even made to prohibit its study at the University of Paris. However, St. Thomas Aquinas (*d.* 1274) harmonized as much as possible of the doctrines of Aristotle and his Muslim commentators with Christian theology to form a natural philosophy, based squarely upon the geocentric doctrine, with which Copernicus and the other pioneers of modern science had later to contend.

Following the literary discovery of the astronomical systems of Aristotle, of Ptolemy, and of al-Bitrugi, the old controversy over the status of planetary hypotheses and over the rival claims of these systems to represent physical reality was renewed in the schools of the West. It figures largely in the writings of the great Scholastic Doctors of the thirteenth century—Albertus Magnus, who favored the Ptolemaic theory, and Roger Bacon and St. Thomas Aquinas, who admitted that the system of eccentrics and epicycles agreed best with the facts of observation but found it incompatible with the physical principles that alone satisfied the reason, and who awaited some solution of the problem that should be acceptable in all respects. The practice became established of conceiving the universe ideally as a system of concentric spheres, while representing the planetary motions in numerical detail on Ptolemaic lines by means of geometrical devices or of equivalent material mechanisms constraining the motions of the planets. (In Italy, however, a last attempt was made, by Fracastoro, less than ten years before the publication of Copernicus' book, to represent the planetary motions by means of an elaborate combination of spheres.)

Meanwhile, Aristotle's whole theory of science was being subjected, within the Schools, to criticism resulting in an extension of the roles of observation and experiment and in a greater concentration upon the mathematically specifiable aspects of experience. This movement, supported by progress in technology, led eventually to the scientific revolution of the seventeenth century.

As early as the fourteenth century there occurred in Paris a revolution in ideas concerning the motion of projectiles that marked an important step toward the establishment of a science of mechanics applicable to celestial and terrestrial bodies alike.

It had been a fundamental doctrine in the School of Aristotle that a projectile could be kept in flight only by the direct and continuing application of a motive power thereto. Since a projectile, once launched, loses contact with the projecting agent, it was necessary to assume that the act of projection imparted an impulse to the surrounding air, which in turn impelled the projectile and also transmitted the impulse to the adjacent layer of air, and so on. Thus the motion was maintained until the impulse became too weak to suspend the action of gravity any longer, and the projectile fell to the ground. This view of the matter was questioned, as early as the sixth century A.D., by John of Alexandria, called Philoponos; he preferred to conceive the act of projection as imparting an incorporeal motive power to the projectile, the subsequent motion of which was hindered rather than promoted by the presence of the medium. The more critical-minded Schoolmen had no difficulty in demonstrating the absurdity of Aristotle's theory of projectiles. In the fourteenth century Jean Buridan, of Paris, introduced the important conception of *impetus,* a motive power or quality conceived to be conferred by the act of projection upon the projectile and jointly proportional to the weight and the speed of the latter. It was this impetus that maintained the projectile in flight, and it was its progressive weakening by gravity or air resistance that eventually brought the motion to an end. Buridan conceived each heavenly sphere as moved not by an angelic Intelligence but by an impetus conferred upon it by God at the Creation. Buridan's ideas marked a step toward the seventeenth-century doctrine of inertia.

The sounder mechanical ideas of the Parisian School suggested the impossibility of deciding, on the strength merely of considerations drawn from experience, whether the diurnal motion belonged to the earth or to the heavens. One of Buridan's disciples, Nicole Oresme, sought to refute the arguments against the rotation of the earth in a commentary attached to his French translation (1377) of Aristotle's *De caelo et mundo.* Some of Oresme's arguments strangely anticipated those of Copernicus (see Chapter III, § 3, *infra*); the latter, however, is unlikely to have had direct access to the commentary, which was never printed.

The disorders of the Hundred Years' War and the Papal Schism brought about the eclipse of Parisian science; in the fifteenth century, however, there occurred a remarkable revival of astronomical studies, associated chiefly with the Italian and the German universities, which was part of the wider movement of the Renaissance. It was stimulated by the recovery of the classics of ancient science and philosophy in their original Greek instead of in corrupt translations. The widespread study of these works and the creation of a secondary literature inspired by them were of course favored by the invention of printing. With the widening of the literary horizon, opinions which Aristotle had ignored or flouted now came again to be discussed—the theories of Philolaus, of Heraclides, of Aristarchus, of the Atomists with their vision of an infinite universe, all tending to bring the geocentric theory into dispute. Other factors helping to break down age-long intellectual inhibitions were the rise of nationalism in Europe and the development of the reform movement in the Church. Practical motives for the improvement of astronomical knowledge were supplied at that period by the need for reforming the calendar (which had fallen into great disorder with the passing of the centuries) and by the demands of ocean navigators for the nautical instruments and tables which they required on their voyages of competitive exploration and trade.

Such were the principal historic factors bearing upon the state of cosmological speculation when, in the sixteenth century, planetary theory was advanced by the achievements of Nicolaus Copernicus.

2

The Life Story of Copernicus

BEFORE DISCUSSING THE CONTENTS OF THE HISTORIC BOOK WHICH was to bring the age-long supremacy of the Ptolemaic system to an end, we shall endeavor to give some account of the life of its author. The generations immediately following Copernicus allowed many precious memorials of him to perish; and the reconstruction of the great astronomer's career and intellectual development effected by the laborious researches of the past century still is not complete. For the substance of the following biographical sketch we shall rely principally upon the elaborate studies of L. Prowe, L. A. Birkenmajer, and Ernst Zinner.

§ 1. BIRTH AND PARENTAGE

The astronomer whom we know as Nicolaus Copernicus devised that appellation for himself, in accordance with the scholarly custom of his day, by Latinizing both his Christian name, Niklas, and the family surname, which was variously spelled in contemporary records but which seems to occur most frequently in the form Koppernigk. He employed the Latin form only in his learned publications and occasionally in letters and inscriptions; for ordinary official purposes he wrote his name Coppernic. He usually doubled the p; and this spelling is therefore regarded by some as the correct one. But in the last few years of his life he seems to have preferred to spell his Latin name with a single p; this is the spelling found in the *De revolution-*

ibus and preserved through all the editions of that book; it is, moreover, that which has usually been adopted by English writers, and it will be retained in the present work.

The family name of the Koppernigks has a bearing upon the question of their extraction, or their nationality, so far as that term is appropriate to the late Middle Ages. This question has been a matter of keen controversy between German and Polish writers, each anxious to claim the great astronomer as a fellow countryman. It appears that the cradle of the family was a town called Koppernigk, near Neisse, in Silesia; it is mentioned in extant documents from the middle of the thirteenth century onward (G. I. Bender, *Heimat und Volkstum der Familie Koppernigk* [*Coppernicus*], Breslau, 1920; *Darstellungen und Quellen zur schlesischen Geschichte,* Band 27). The name may have been Slavic in origin; but German scholars have been at pains to point out that as we know it (with three syllables and the doubled *p*) it is German. There is documentary evidence that Koppernigk was inhabited by Germans at the period when migrants were moving eastward from Silesia into Poland bearing the names of their places of origin. By 1400, people called Koppernigk were prominent in the affairs of the principal Polish cities, their frequent choice of Niklas as a Christian name for their boys being connected, perhaps with the fact that St. Nicholas was the patron saint of the parish church of the old home town in Silesia.

The economy of Poland was expanding rapidly at the end of the fourteenth century. A marriage alliance with Lithuania in 1386 had enlarged the territory and secured the eastern flank of the Poles, who, under the royal house of the Jagellons, were in a position to dominate eastern Europe and to check the political expansion of the Germans into that area. Adventurous immigrants from the west were, however, encouraged to settle in the Polish cities, where they formed a prosperous bourgeoisie filling the social gulf between the nobility and the serfs.

In the municipal records of the Polish capital, Cracow, for the early fifteenth century, mention is made of one Johann Koppernigk, a merchant and banker. In due time his name gives place to that of a Niklas Koppernigk, who had much the same business relations as Johann and was probably his son; he was to become the father of the great astronomer. At some time not later than 1458 this Niklas Koppernigk migrated with many of his associates from Cracow to Torun, on the Vistula. Torun (Plate I), which is now in Poland, had been founded in the thirteenth century by the Teutonic Knights to form an outpost of the independent state which they had carved out for themselves by their conquest of the heathen Prussians. During the fourteenth century Torun had flourished as a port of the Hanseatic League and an entrepôt for the trade of western Europe with Poland; but by the end of the fifteenth century the town had lost much of its former prosperity through the rivalry of Danzig and the frequent hostilities between Poland and the Teutonic Knights. Soon after the arrival of Niklas Koppernigk in Torun, the rebellious subjects of the now decadent Order made common cause with the Poles; the Knights lost their independence and much of their territory, and Torun passed, with West Prussia, under the suzerainty of the King of Poland. The migration of Niklas Koppernigk from Cracow to Torun may have reflected the worsening of relations between Poles and Germans in the Polish capital.

At Torun, Niklas Koppernigk prospered in business and was appointed a magistrate for life. Not later than 1464 he married Barbara Watzenrode (or Watzelrode), the daughter of a wealthy Torun merchant; hers was an established German family from which the town had drawn many of its councilors and magistrates, and which had suffered for its sympathies with the Teutonic Knights in the recent war. There were four children of the marriage, of whom Niklas, the future astronomer, was the youngest. He was born on February 19, 1473, in a house in

Torun, that is still identified by tradition. He had a brother, Andreas, later the companion of his foreign travels and studies, and two sisters—Barbara, who became a nun, and Katherina, who married a merchant of Cracow. The astronomer's father and mother both belonged to a class drawn exclusively from the German elements of the population, and such indications as he gives of his own affinities suggest that he identified himself with the Germanized West Prussians, as distinguished on the one hand from the Poles and on the other from the Teutonic Knights. Such distinctions were still softened, however, by the prevalence of a universal faith and a common learned language; militant nationalism and the myth of race belong to a later Europe than that in which Copernicus spent his days.

§ 2. YOUTHFUL STUDIES AND TRAVELS

When Niklas was ten years of age his father died, and the children were adopted by their maternal uncle, Lucas Watzenrode (1447–1512); he was a man of strong character, who, after a distinguished academic career in Italy, had entered the service of the Church and was now well on his way to a bishopric. Lucas sent his nephew to school at Torun and later at Wloclawek, some way up the Vistula, and thence, in 1491, to the University of Cracow, his own alma mater, which, under royal patronage, had acquired a reputation second to none among the northern universities. A brilliant school of mathematics and astronomy had been built up at Cracow by Albert Brudzewski, with whose works Niklas would become acquainted even though the two men seem never to have met. Although adhering to the Ptolemaic system in his published works, Brudzewski was a man of liberal sympathies. He was on friendly terms with leaders of the new humanistic movement, which, emanating from Italy, had already begun at Cracow to challenge the traditional Scholastic discipline and to produce cleavages in the university that led occasionally to street fighting between rival student factions. Niklas was enrolled, like other freshmen, in the Faculty of Arts, which provided courses in classics, theology, philosophy, mathematics, and so forth, introductory to the more specialized faculties, such as divinity and medicine. It was probably at the Polish university that he became accustomed to the use of astronomical instruments and to the practice of observing the heavens, and it was probably there that he took the first steps toward the construction of his new system of astronomy. At Cracow, too, surrounded by students from many parts of Europe, Niklas may well have acquired that command of Latin to which his later writings bear witness, and may have taken the Latin name by which he has ever since been known and by which we shall designate him in the following pages.

After spending probably three years at Cracow, Copernicus returned home. His uncle had been since 1489 the Bishop of Ermland (or Varmia), one of the four dioceses into which Prussia had been divided; it formed a little principality enjoying a large measure of independence. The Bishop's palace was situated at Heilsberg (Plate II) and the cathedral was at Frauenburg, on the coast. Bishop Lucas was eager to provide for his nephew's future by having him elected to a canonry of Frauenburg Cathedral. The first attempt proved unsuccessful, and the young astronomer was given permission to resume his university studies, this time in Italy. Setting out in 1496, he made his way across the Alps to Bologna, and for the next four years he studied there in the school of law for which the city had long been famous. In 1498 he was joined by his brother, Andreas, who had been with him at Cracow. They were both enrolled in the *Natio Germanorum*—the most considerable of all the "nations" into which foreign students were organized at Bologna. The two young men seem to have joined freely in the student life of the city, and upon at least one occasion they were obliged to appeal urgently for funds to an emissary from Ermland, who happened to be within reach.

While at Bologna, Copernicus came into close personal touch with Domenico Maria da Novara (1454–1504), the professor of astronomy there; it is possible that he even lodged in Novara's house. Novara was a brilliant teacher and a critical-minded observer. He believed (mistakenly) that he had discovered a systematic increase, since the time of Ptolemy, in the latitudes of several places in southern Europe, which he attributed to a progressive displacement of the pole of the heavens. He was also one of those who detected the diminution that had occurred in the obliquity of the ecliptic since ancient times. These considerations disposed Novara to be rather dubious about the accepted system of astronomy. Moreover, he was one of the leaders in a revival of Platonism which was just then sweeping over southern Europe. In the spirit of this movement he would strive to conceive the constitution of the universe in terms of simple mathematical relations; and unfettered converse with such a man must have encouraged Copernicus in any plans which he might already have framed for the reform of astronomy along similar lines. Novara and Copernicus observed the heavens together when they had opportunity, the younger man being, according to Rheticus, "not so much a pupil as a helper and witness of the observations" (Narratio prima).

The earliest observation of his own which Copernicus utilized explicitly in his book (an occultation of Aldebaran by the moon, De rev., IV, 27) belongs to this Bologna period. It has now been established that Copernicus took his M.A. degree at Bologna. In the spring of 1500 he went to Rome to be present at the Easter celebrations of that great Jubilee Year, and he remained for a whole year in the city, teaching mathematics privately; to this period, also, belongs an observation of a lunar eclipse which he subsequently employed in constructing his lunar theory (De rev., IV, 14).

About 1497 Copernicus had been elected a canon of Frauenburg in his absence; his brother had obtained similar prefer-
ment in 1499. Like many ecclesiastics of his day, Copernicus
appears to have entered the service of the Church from temporal motives, and seems never to have proceeded beyond the
vows necessary for his admission to the Chapter. In the summer
of 1501 the two young men returned to Ermland to request further leave of absence, so that they might continue their studies
in Italy. This was granted, and Copernicus set out for Padua,
where he completed his legal studies. It was at about this period,
and probably at Padua, that Copernicus learned Greek; he
thereby gained direct access to the works of Plato and to the
other Greek writings from which he later claimed to have derived inspiration. Copernicus was not graduated at Padua, but
went on to Ferrara to take his doctorate in canon law in 1503.
Returning to Padua, he embarked on the study of medicine,
which in those days was taught mostly from the works of authorities such as Galen and Avicenna, with occasional dissec-
tions for illustration. It was not unusual, at that period, for a
churchman to learn something of the healing art, although he
was expected to eschew surgery. Copernicus probably was not
graduated in medicine, and by the beginning of 1506, at latest,
his years of study abroad were ended and he was back in Ermland.

Just as Copernicus was at last settling down to his official
duties at Frauenburg, however, he was bidden to take up his
residence at the Bishop's palace, some forty miles away, to act
as medical adviser to his uncle, whose health was uncertain. His
next six years, until Lucas Watzenrode's death, in 1512, were
accordingly spent in the stately castle of Heilsberg (Plate III).
It was most probably here that Copernicus began to give a definite literary form to the new system of cosmology the elaboration of which was to occupy him during the remaining thirty
years of his life. The first fruit of this activity was the short

synopsis of the new planetary system, in its earlier form, given by Copernicus in his manuscript *Commentariolus* (see Note III, *infra*), if we follow Birkenmajer in assigning its composition to 1512 or earlier, as against the much later date previously adopted by M. Curtze.

It was during these years at Heilsberg, too, that Copernicus became acquainted at first hand with the intricacies of Ermland's politics. The little principality was hard put to maintain its independence against two overbearing and mutually hostile neighbors—Poland and the Teutonic Knights (whose territory surrounded it on three sides). To the Bishop fell also the delicate task of mediating between the German population of West Prussia, which aspired to complete national freedom, and the Poles, who were bent upon incorporating the province in their kingdom. Copernicus accompanied, or represented, his uncle on a number of diplomatic missions. While staying at Cracow in 1509, he took the opportunity to have published his own Latin translation of the Greek Epistles of the Byzantine poet Theophylactus Simocatta. This version, which Copernicus dedicated to his uncle, was the only book he ever had published other than his great work on astronomy. It was on the return journey from another visit of the Bishop and his nephew to Cracow, three years later, that Lucas Watzenrode was overtaken by mortal illness; he was carried to Torun, his birthplace, and there he died, at the end of March 1512. With the death of his protector, Copernicus' attendance at Heilsberg came to an end, and he returned to Frauenburg. In later years, however, he was frequently summoned to the palace to give medical advice and treatment to the elderly and ailing Bishops who successively occupied the see; there is also a credible tradition that the poor of the district were allowed to benefit by his skill. Two years before his death, Copernicus was hastily summoned to Königsberg by the Duke of Prussia to attend one of his councilors who had fallen dangerously ill. He returned to Frauenburg after about a month's absence, leaving the patient out of danger. A number of Copernicus' medical books have been preserved, together with recipes written in his own hand in margins and on flyleaves; they give the impression that in matters of healing Copernicus followed the accepted authorities, and applied the customary remedies, of his day.

§ 3. THE CANON OF FRAUENBURG

Frauenburg Cathedral, about which the life of Copernicus was henceforward to revolve, stands on a low hill overlooking the Frisches Haff, a fresh-water lagoon opening into the Gulf of Danzig (Plate IV). It was fortified against the perils of those disordered times by a surrounding wall; and a long-standing tradition claims the three-storied tower which forms the north-west corner of the enclosure as the astronomer's old abode. Several of his new companions were natives of Torun, who would be personally known to him. He was tragically deprived, however, of the society of his brother, who, shortly after his return from Italy, developed an incurable disease, probably leprosy, and was forced to retire from his cathedral duties; after vainly seeking relief abroad, he died, not later than 1519.

It was the function of the canons of Frauenburg, who numbered about sixteen, to carry on the services of the sanctuary and to assist the Bishop in the spiritual and temporal administration of the diocese. They all lived within the cathedral enclosure. Their income was derived from real estate, and they were permitted to maintain an establishment of servants and horses. It was not difficult for them to obtain leave of absence from duty, as Copernicus frequently did.

It was at Frauenburg that Copernicus made nearly all those observations of his own which he utilized in the numerical determination of the elements of his planetary theory. His book contains twenty-seven of his own recorded observations; but if to these we add those written in such places as the flyleaves and margins of books, the total amounts to more than double that number. Of Copernicus' activities and resources as an observer, relatively little is known. His instrumental equipment at Frauenburg seems to have been of a modest order. The instruments the construction and the use of which he describes in his book are those of Ptolemy and his successors.

There is, in the first place, an account of a device for determining the meridian altitude of the sun (*De rev.*, II, 2; cf. *Alm.*, I, 10). It consisted of a square slab of stone or metal, with one of its faces made exactly plane. Upon this face, with one of the upper corners as center, there was engraved a quadrant of a circle, which was divided into degrees and minutes of arc. A short cylindrical style was attached at the geometrical center of the quadrant and at right angles to its plane. The quadrant was set up in the plane of the meridian, and the shadow of the style cast at noon upon the graduated arc indicated the sun's meridian altitude. This instrument was suited to determine the inclination of the ecliptic to the celestial equator—an angle given by half the difference between the meridian altitudes of the sun at the summer and winter solstices respectively. Copernicus seems to have possessed such an instrument and to have employed it to determine this angle, which he found to have suffered a diminution since Ptolemy's day (*De rev.*, II, 2).

Copernicus also explains the construction and use of the Ptolemaic astrolabe, or armillary sphere (*De rev.*, II, 14; cf. *Alm.*, V, 1); this was a combination of concentric metal rings which were made respectively to coincide with the planes of the principal circles of the celestial sphere. The direction of a celestial body could be referred to graduations on these rings with the aid of movable indices, and its celestial coordinates (longitude and latitude) could thus be determined. The instrument chiefly employed by Copernicus in his own observations, however, was of the type known as a *triquetrum,* which he called, after Ptolemy, an *instrumentum parallacticum* (*De rev.*, IV, 15; cf. *Alm.*, V, 12); he employed it to measure altitudes of the heavenly bodies. This instrument, which, according to Gassendi, he constructed with his own hands, consisted of three graduated pine-wood rulers (Fig. 5); one of these, AB, was fixed in a ver-

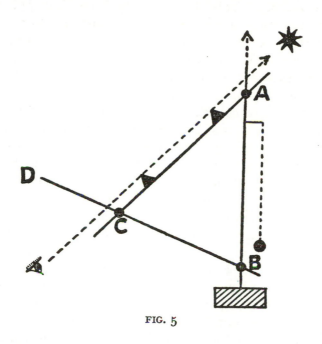

FIG. 5

tical position, and it had, at its upper end, a pin, A, about which the second ruler, AC, was free to turn in a vertical plane. This second ruler carried two sights, and it could be directed accurately toward any celestial body the angular distance of which from the zenith (or elevation above the horizon) was required. On these two rulers, and at equal distances from the pin A, were two other pins, B and C. About B the third graduated ruler, BD, was free to turn; it was longer than the other two, and at each division it had a perforation into which the pin C could be inserted. It thus served as a crosspiece to hold the two rulers AB and AC at any required angle with each other. The length of this crosspiece intercepted between B and C was the chord of the angle BAC which the line of sight CA made with the vertical; this angle could be deduced from the length of BC with the aid of a Table of Chords (see Note II, *infra*); it gave the zenith-distance of the object toward which the line of sights was directed. To facilitate the calculation, the lengths AB and AC were each divided into 1000 units, and the crosspiece was made long enough to contain at least 1414 ($= 1000 \sqrt{2}$) of these units, so that all values of the angle BAC up to 90° could be measured, to a certain degree of approximation. The trique-trum of Copernicus was preserved at Frauenburg for some forty years after the astronomer's death; it was then given to Tycho Brahe, who valued it highly; but its fate is unknown (see J. L. E. Dreyer: *Tycho Brahe*, Edinburgh, 1890, 103, 125).

In 1514 Copernicus was invited by the Lateran Council to assist in a proposed attempt to reform the calendar, which had become deranged with the lapse of centuries, partly through the overestimate of the mean length of the civil year in the Julian reform, and partly through the inaccuracy of the relation assumed to hold between the lengths of the lunar month and the tropical year. Copernicus pointed out that any attempt at reform would be nugatory unless the motions of the sun and moon were first precisely ascertained; but he promised to keep the problem in mind. He refers to it at the end of the Preface of his book (1543) as partly justifying his efforts to refashion astronomy, and, indeed, the improved tables based upon the Copernican theory paved the way for the reform of the calendar introduced by Pope Gregory XIII in 1582. This connection of the researches of Copernicus with the practical problem of re-forming the calendar may be of interest to those who insist upon finding some utilitarian motive behind every advance of science.

During the early years at Frauenburg, his part in the ordinary duties of the little community left Copernicus ample leisure for pursuits of a more philosophical character. But his wide experience of the world, and the knowledge of affairs which he had acquired while in attendance upon his uncle, marked the young canon for duties of especial responsibility. Accordingly, in November 1516, he was appointed to administer the temporal and spiritual affairs of some outlying estates belonging to the Chapter. He held this commission for three years and subsequently for a further half year (1520–21).

During this period Copernicus lived at Allenstein Castle (Plate V), visiting Frauenburg only occasionally to see his old friends and to make isolated celestial observations. His term of office fell at a time of growing difficulty and danger for the whole of Ermland. The war clouds were gathering over Poland and East Prussia; they broke, at the end of 1519, in a campaign of pillage in which the little principality suffered severely from the plundering bands of the Teutonic Knights. Heilsberg was bombarded; Frauenburg had to repel a raid; and Copernicus, at Allenstein, was for a time threatened by the forces of the Order. The hour of crisis revealed remarkable qualities of leadership and resourcefulness in Copernicus. He had to undertake exceptional responsibilities during the dispersal of the Chapter to various places of refuge; when the armistice of 1521 had brought large-scale hostilities to an end, he was active in resettling the deserted Allenstein estates; and it fell to him to draw up a memorial on the wrongs suffered by Ermland in the war and to present it at the peace conference.

About this time the debasement of the Prussian coinage, which had been aggravated by the war, was giving rise to much inconvenience and hardship. Copernicus gave this matter very earnest attention. His analysis of the causes and evil consequences of such debasement and his recommendations for remedying it were set forth in 1522 in a memorandum, written in

German, which he subsequently revised and drew up in Latin for presentation to the Prussian *Landtag* of 1528. He urged that the minting of coins should be a state monopoly, instead of each city's or each district's having its own currency; that the quantity of money in circulation should be controlled; that in each denomination the coins should contain not less than a certain weight of precious metal (although this might be alloyed to give the coins bulk and hardness); and that the old coinage should be withdrawn when the new was issued, in order to prevent the good new coins from being bought up and melted down. Transitional hardships attending the fulfillment of contracts based on the old currency would have to be sympathetically considered. These recommendations of Copernicus were made the basis of legislation. Under more favorable conditions he might have rendered his country services at the Mint as distinguished as those of Newton; but the obstructive tactics of interested powers prevented any effective action for the time being.

About the same period Copernicus circulated copies of an open letter to his friend Bernhard Wapowski, severely criticizing the speculations of a certain Nuremberg astronomer, Johann Werner, who had sought to revive the old hypothesis that the equinoctial points oscillate slowly about their mean positions (see Chapter IV, § 2, *infra*).

In 1523 Copernicus was appointed Administrator-General of the diocese during a six months' interregnum between two Bishops. Thereafter, however, his more arduous duties and responsibilities gradually passed into the hands of younger men. By now the political background had become less alarming, the Grand Master of the Teutonic Knights having agreed to become a hereditary secular duke under the suzerainty of the King of Poland. On the other hand, the astronomer's later years were troubled somewhat by an unsympathetic Bishop, by the domestic dissensions of the Chapter and its disputes with the King of Poland over the right to elect Bishops, and especially by the unrest and cleavage accompanying the spread of Lutheran doctrines from Germany and the attempts to repress these doctrines. Judging by the tone of a polemical book published in 1525, at Copernicus' instigation, by his old friend Tiedemann Giese, a fellow member of the Chapter, Copernicus seems to have been orthodox in his opposition to Luther but eager to resolve the conflict in a conciliatory spirit and to avoid the disruption of the Church.

In the spring of 1539 Copernicus received an unexpected visit from a young German scholar, Georg Joachim von Lauchen (1514–76), who had adopted the appellation Rheticus after the old Roman name (Rhaetia) of the district of his birth. Rheticus was a protégé of Melanchthon. Although only twenty-five years old, he was already a professor of mathematics at the Protestant University of Wittenberg. He explained that his interest had been awakened by what he had heard of the doctrines of Copernicus, and that a keen desire to know more about them had brought him all the way to Frauenburg. Rheticus was received cordially by the astronomer, who gave him every assistance in mastering the intricacies of the new system of cosmology, both by permitting him to consult the manuscript in which it was set forth and by way of personal explanation.

Rheticus had promised to send details of the Copernican system to his old teacher, Johann Schöner, the Nuremberg astronomer, should his mission to Frauenburg succeed. After some ten weeks of study and discussion of Copernicus' manuscript, Rheticus drew up an account of its contents and addressed it to Schöner; it was published, with the approval of Copernicus, at Danzig, in 1540, under the title *Narratio prima de libris revolutionum*. This was the earliest explicit account of the Copernican system to be published. (For a short summary of its contents, see Note III, *infra*).

Rheticus spent more than two years at Frauenburg or traveling about the district. He made many acquaintances, among them Copernicus' old and trusted friend Tiedemann Giese, now Bishop of Kulm. The prolonged visit of Rheticus to Copernicus was not without peril to both. For Rheticus had come from a stronghold of Protestantism into the jurisdiction of a Bishop notorious for his stern measures against heretics, and he had been welcomed by a man whose opinions were at that time more obnoxious to the Wittenberg authorities than to orthodox Catholics. Nevertheless, Rheticus lingered in Prussia until the autumn of 1541, when he left to resume his duties at Wittenberg. He took back with him a transcript of the two chapters of the *De revolutionibus* (I, 13 and 14) which deal with the elements of plane and spherical trigonometry (mostly collected and generalized from the *Almagest*), and he published them (separately from the main work) at Wittenberg in 1542, under the title *De lateribus et angulis triangulorum tum planorum rectilineorum, tum sphaericorum libellus*, etc. Rheticus further sought to express his gratitude to his host and instructor by the gift of a number of recently printed books on mathematics and astronomy; these included the first Greek edition of Ptolemy's *Almagest* (Basle, 1538). They formed a notable addition to the little private library which Copernicus bequeathed to the Cathedral at his death; these books were later removed to Sweden by Gustavus Adolphus, and many of them are now preserved at Upsala.

In the Preface to his great work of 1543, Copernicus alludes to the urgent exhortations of Tiedemann Giese and other friends that he publish the manuscript which he had kept under periodical revision for some thirty years (*"in quartum novennium"*). The gist of his teachings had become generally known among scholars through the circulation of the *Commentariolus*. This little tract probably served to provide the material for a lecture on the Copernican system which was delivered in 1533 to Pope Clement VII and his court by the papal secretary, Johann Widmanstad, in the gardens of the Vatican. This lecture prompted Cardinal Schönberg, who, as nuncio in Poland and Prussia, had met Copernicus years before, to write to the astronomer from Rome in 1536, strongly urging him to make the full details known to the learned world and, in any case, to send particulars to the Cardinal himself, at the latter's expense. Copernicus had reason to be well pleased with the laudatory tone of this letter, since it came from such a quarter; it appears in the place of honor in his book. (Fame of a different sort came to him through a satirical play staged at Elbing, a few miles from Frauenburg, which poked fun at the star-gazing ecclesiastic and his extraordinary opinions.) Moved by the persistent entreaties of his friends, by the youthful enthusiasm of Rheticus (whose *Narratio* had fulfilled a useful preparatory purpose), and perhaps by the sense that his own life was drawing toward its close, Copernicus at length consented to the publication of his book. He entrusted the manuscript to Giese, who sent it to Rheticus, probably in accordance with a provisional arrangement already made during the young Protestant's visit to Prussia. The work was printed at Nuremberg and was published early in 1543 under the title *De revolutionibus orbium coelestium libri VI.*

During the winter Copernicus had been overtaken by serious illness. A paralytic stroke supervened, leaving no hope of his recovery. For weeks he lay awaiting death. At length, on May 24, 1543, an advance copy of the newly published book was brought to him. He saw and handled his completed work, and some hours later he passed away.

Rheticus is reported to have compiled a biography of Copernicus; but neither this nor any other contemporary account of the astronomer's life has come down to us, and almost all his letters have disappeared. Only portraits, of various degrees of authenticity, remain to give us some idea of the man himself; they reveal a grave and thoughtful ecclesiastic of somewhat severe countenance. Copernicus fulfilled admirably the Renais-

sance ideal of universal scholarship. In the field of astronomy
he had to overcome formidable difficulties even in such matters
as the chronology of ancient observations dated with reference
to calendars the relation of which to the Christian reckoning
was obscure. He had also to spend in tedious computation hours
which would now be shortened by the use of tables and calculating machines. It is with Copernicus' contributions to astronomy
that we are concerned almost exclusively in these pages. However, he was also an accomplished mathematician, the trigonometrical chapters of his book being thought worthy of independent publication by Rheticus. He was a linguist with a command of Polish, German and Latin, and he possessed also a
knowledge of Greek rare at that period in northeastern Europe
and probably some acquaintance with Italian and Hebrew. We
have referred to his activities as a physician and to his essays in
the theory of currency. Copernicus has also enjoyed a reputation
as an artist on the strength of his self-portrait, only a copy of
which survives, attached to the astronomical clock in Strasbourg
Cathedral. His strength of character and tenacity of purpose are
written large in the great book which he brought to completion
in the face of foolish prejudice and learned opposition.

§ 4. THE COMPOSITION AND PUBLICATION OF THE *De Revolutionibus*

Although the external facts concerning the life history of
Copernicus can be traced in some detail, we are almost entirely
without information as to the motives and influences which
may have prompted him to undertake the reformation of astronomy, and as to any transitional stages through which his planetary theory may have passed before it reached the form in which
it is presented in the *De revolutionibus*. The elaborate investigations of L. A. Birkenmajer (see Bibliography) have, however,
thrown much light, both upon the sources of information which

Copernicus utilized and upon the gradual development of his technique.

Birkenmajer has made a critical examination of the original manuscript of the *De revolutionibus* and of the notes written in Copernicus' own hand which have been found in some of the books formerly in his possession, as well as of other relevant material. From the results of his researches it appears that Copernicus acquired his copies of several fundamental works on mathematics and astronomy while he was still a student at Cracow (*c.* 1491–94); these included the *Tabulae directionum* of Regiomontanus and the *Alfonsine Tables*. The manuscript notes of Copernicus in the latter book appear to date back to the Cracow period of his career, and they include calculations apparently relating to the heliocentric planetary scheme presented in the *Commentariolus*. This would seem to suggest that Copernicus took the first steps in the construction of his system some time before his departure for Italy. The numerical data for the *Commentariolus* seem all to have been derived from the *Alfonsine Tables*. The chief sources of information employed in the composition of the *De revolutionibus,* however, were the *Epitome in almagestum* (1496) of Purbach and Regiomontanus and the Latin translation of the *Almagest* by Gerard of Cremona (Chapter I, § 2, *supra*), published at Venice in 1515. The Greek edition of the *Almagest* (Basle, 1538), which Copernicus received as a present from Rheticus shortly before his death, came too late to be of any use to him in the elaboration of his system.

The original manuscript of the *De revolutionibus* affords evidence that the heliocentric theory underwent a gradual development in the mind of Copernicus as regards its precise form; moreover, it warrants inferences as to what the stages of this development were and when they were reached. During the three centuries following the publication of the book, the manuscript passed from hand to hand; it ceased to be recognized for what it was, and all trace of it had been lost when, in the middle of the nineteenth century, it was discovered in a nobleman's library at Prague. The manuscript is not simply a fair copy written out after Copernicus had put the last touches to his masterpiece; it is a heterogeneous document, embodying numerous and extensive alterations, insertions, and cancellations, evidently made at various dates. It has been the aim of critics to analyze the final text and the suppressed passages into successive layers, as it were, and to assign to each of these an approximate period of composition, based upon the dated observations embodied therein or upon other, more indirect, evidence.

The manuscript was examined by Curtze for the purpose of establishing the text of the 1873 edition; later it was scrutinized more minutely by Birkenmajer and A. Czuczynski, who took account of the variations in the characteristics of the handwriting and in the quality of the paper and ink employed. Curtze formed the opinion that the manuscript had undergone two successive recensions. The second of these was presumably not completed before 1529, as the final version contained particulars of an observation made in that year (see Introduction to edition of 1873). Birkenmajer assigned the two revisions of the manuscript to the periods 1515–19 and 1523–32, respectively; and he thought that probably some finishing touches were applied about 1540, during Rheticus' visit. Copernicus did not cease to observe the heavens when his manuscript was substantially completed, about 1530; the series of his recorded observations extends down to 1541. He may well have felt that the numerical constants of his theory left much to be desired. Among the canceled passages of the manuscript Birkenmajer found traces of a geometrical theory of the planetary motions in longitude, differing slightly from that set forth in the published work. It evidently corresponded to an early stage in the elaboration of the Copernican system; and it is of interest to note that it is this earlier form of the theory that is outlined in the *Commentariolus* (see Note III, *infra*). Birkenmajer was convinced that the editors of 1873 had misread the manuscript in a number of places and, further, that some of the textual alterations previously attributed to Copernicus had in fact been made by another hand—in time, however, for them to be followed by the printers of the 1543 edition. He also compared the manuscript of the *De revolutionibus* with that of the unpublished and long-lost commentary on Copernicus by Erasmus Reinhold, which he discovered at Berlin; he came to the conclusion that Reinhold must have seen Copernicus' manuscript before it went to press and that it was probably he who interfered with the text. There are also remarkable discrepancies between the manuscript and the printed edition of 1543. The manuscript, which nowhere bears the name of its author, and which is untitled, doubtless called for some editorial attention before it could be sent to press. The divergences of the printed text from the manuscript are, however, so considerable and arbitrary as to suggest to the editors of 1873 that the Nuremberg printers did not work from the manuscript but from a transcript made by someone acquainted with astronomy, who sought to improve the style and who did not scruple to make unauthorized additions and omissions. It may be that Reinhold was the scribe. There were, however, other exceptional circumstances attending the first appearance of the *De revolutionibus,* as we must now relate.

The task of printing and publishing Copernicus' book was entrusted by Rheticus to his friend Johann Petrejus, of Nuremberg. Rheticus seems to have intended to see the book through the press himself, but before the work had progressed very far he was obliged to leave Nuremberg for Leipzig. He handed over the task of supervision to Andreas Osiander, a local Lutheran theologian and mathematician of some note. Osiander had previously had some correspondence with Copernicus, portions of which were subsequently published by Kepler. It seems that Copernicus inquired (July 1, 1540) whether it would be possible for him to publish his theory of the earth's motion without exciting hostile criticism. Osiander replied (April 20, 1541): "For my part, I have always felt about hypotheses that they are not articles of faith, but bases of calculation, so that, even if they be false, it matters not so long as they exactly represent the phenomena of the [celestial] motions. . . . It would therefore seem an excellent thing for you to touch a little on this point in the Preface. For you would thus render more complacent the Aristotelians and theologians whose contradiction you fear." (Kepler's *Apologia Tychonis contra Ursum.*) Copernicus, who had a very different notion of an astronomical theory (see Chapter III, § 5, *infra*), ignored this advice. But, as we have seen, a train of circumstances placed Osiander in control of the final stages of the publication of the book; and he took advantage of the situation to insert, in the most prominent position, a brief Preface such as he had counseled Copernicus to prepare. Scholars will doubtless be shocked (writes Osiander) by the unsettling hypothesis of the earth's motion set forth in this book. They should remember, however, that the astronomer is not concerned with the true causes of celestial motions. He is therefore free to adopt any hypothesis which may enable him to give a geometrical representation of the motions that have been observed in the past, and to predict the motions that will occur in the future. Such a hypothesis need not be true or even probable; it is sufficient that it should lead to results in agreement with the facts of observation, and that it should be the simplest hypothesis capable of so doing.

Osiander's *Praefatiuncula,* as Gassendi called it, probably represented a well-meaning effort to disarm criticism and to ensure a favorable reception for the book; and it seems to have succeeded in its object. It is anonymous, but it can scarcely be called a forgery, since it does not purport to be by Copernicus, to whom it refers in the third person and in laudatory terms. Its authorship early became known to one or two, but was first revealed to the learned world in general by Kepler (*Apologia Tychonis contra Ursum*, I, ed. Frisch, Vol. I, p. 245; also *Astronomia nova: Auctor Ramo*, on the verso of the title page, ed. Frisch, III, 136). To Osiander also would seem to be due the "blurb" on the title page of the *De revolutionibus*, bidding the public "buy, read, and enjoy" the book.

As has already been stated, the manuscript which Copernicus handed over to his friends for publication was untitled. It has never been conclusively established whether the title borne by the printed volume was of his own choosing or whether this too represents an unauthorized addition or perhaps an amendment of his choice. Several of his contemporaries erased the words *orbium coelestium* from the title in their copies, thereby giving the impression that the authentic superscription should be simply *De revolutionibus*. It seems probable that Copernicus may have happened upon this phrase in George Valla's translation (1501) of the *Hypotyposis* of Proclus, where the words represent the Latin rendering of the title of a lost astronomical treatise by Sosigenes. The suggestion that the words *orbium coelestium* were an unwarranted addition to the Copernican choice of *De revolutionibus* seems to be obscurely traceable to Rheticus; however that may be, there seems to be no reason for accusing Osiander of tampering with the title. (See E. Rosen, "The Authentic Title of Copernicus's Major Work," *Journal of the History of Ideas*, 1943, iv, 457 ff.)

Of the later editions of the *De revolutionibus*, those of 1566, 1617, and 1854 (see Bibliography) followed the printed text of 1543. The *Säcular-Ausgabe* of 1873, however, which was published in celebration of the fourth centenary of the birth of Copernicus, was based upon a critical study of the original manuscript. Another critical text, edited by F. and C. Zeller, with a facsimile of the manuscript, appeared in 1944, 1949, as two volumes of the new *Gesamtausgabe* of the works of Copernicus.

3

The Mobility of the Earth

THE ANCIENT EASTERN PEOPLES, WHOSE CULTURE FORMED THE matrix of Greek intellectual development, conceived the earth as a stationary platform having a central or otherwise privileged position in the universe. Similarly, in the archaic Greek cosmologies the earth figured as a central, motionless body having the form of a disc, cylindrical frustum, or sphere, which floated on the primeval ocean or hovered in the abyss. And, in fact, it was the conception of the earth as a motionless sphere, poised symmetrically at the center of sphere-shaped space, which the ancients finally embraced, and which they imposed upon medieval thought.

True, there were not wanting in Hellas bold thinkers who deliberately allowed for the possibility of the earth's being neither stationary nor central in space, and who partly realized how the construction of planetary hypotheses might thereby be simplified. Speculative progress along these lines might have led, and indeed, with Aristarchus did momentarily attain, to a completely heliocentric system. But it was prematurely arrested by a combination of factors, among which must be included (*a*) the reluctance of naïve common sense to believe in a motion of the earth not directly perceptible; (*b*) the influence of religious conservatism, eager to claim a unique and privileged status for man's abode, and successively manifested in Greek, Muslim, and Christian circles; (*c*) the growing authority of Aristotle, whose philosophical arguments were solidly in support of the geocentric theory; (*d*) the relative excellence of the planetary tables constructed by Ptolemy and his successors from the standpoint of that theory; and (*e*) the prevalence of astrology, the doctrines of which were inextricably wrapped up with the antique cosmology. Thus it was that the conception of the earth as motionless and centrally situated predominated throughout the Middle Ages. The gradual supersession of the geocentric theory in the sixteenth and seventeenth centuries dates from the appearance of Copernicus' great work of 1543. It will be our task, in this and the three following chapters, to explain what is of most significance in the contents of that remarkable book.

§ 1. THE SCOPE AND PLAN OF THE *De Revolutionibus*

The contents and arrangement of Copernicus' book *De revolutionibus* had better be indicated briefly at the outset. The body of the work falls into six books, each subdivided into a number of chapters. In Book I, Copernicus sets forth his general arguments for believing in the mobility of the earth and for substituting the heliocentric for the geocentric point of view; he sketches the heliocentric arrangement of the solar system in broad outlines; and he gives the modern explanation of the seasons. The book concludes with some elementary plane and spherical trigonometry, including a Table of Chords, or, more precisely, of sines (see Note II, *infra*). Book II deals with spherical astronomy (definitions of the circles of the celestial sphere, transformation of coordinates, etc.) and treats of problems connected with the rising and setting of the sun and of other heavenly bodies. Since the treatment of such diurnal phenomena is independent of rival physical theories as to their causation, no part of this book will again concern us except the star catalogue with which it concludes. Coming now to detailed geometrical schemes of the motions in the solar system, Copernicus treats of the earth's several motions and the elements of its orbit in Book III. In Book IV he deals with the theory of the moon's motions and with the determination of the distances of the sun and moon. In Book V, the longest and the most vital of the six, he investigates the motions in longitude of the five planets and the sizes of their orbits in relation to that of the earth. In Book VI he considers the motions of the planets in latitude.

We shall proceed now to a study of the significant portions of Copernicus' book. In the present chapter we shall examine his grounds for questioning, and his arguments for finally rejecting, the accepted verdict of authority on the matter of the earth's status in the universe.

§ 2. THE APOLOGIA OF COPERNICUS

When Copernicus at last consented to the publication of his manuscript, he took the bold course of dedicating the work to the reigning Pope, the scholarly Paul III. In the Dedicatory Preface he claims for his speculations the Pope's interest and protection, giving some account of how these speculations first took shape. He admits that he has long hesitated to publish his book for fear of the censure which its doctrine of the earth's motion might incur; but his friends have at last prevailed upon him to commit it to the press. The inadequacy of the planetary theories so far proposed, and the great diversity existing among them, must be his chief excuses for adding yet another to their number. The theories which traced their descent from the homocentric spheres of Eudoxus might, he believed, be in accordance with sound physical principles; but there seemed no prospect of their ever affording a precise representation of the planetary phenomena. The rival theories, which had developed from the eccentrics and epicycles of the Alexandrian astronomers, yielded tables of practical value; but they admitted much that was contrary to sound physics, and the universe as conceived on such lines was a monstrosity. Copernicus relates how, in his disappointment at such a condition of things, he turned to the ancient philosophers to see what alternative theories they might have proposed. He found that Cicero attributed a belief in the motion of the earth to one Hicetas, and that similar statements were made by Plutarch about Philolaus, Ecphantus, and Heraclides of Pontus. We have already noted the opinions ascribed to these men in antiquity, and we shall consider later the passages to which Copernicus here alludes and the part that they may have played in the development of his own ideas (see § 6, *infra*). Whether Copernicus really derived his inspiration from these classic passages or whether he merely quoted them for the sake of the impression which they would produce upon his readers, we cannot be certain. At all events he makes them the point of departure for his own exploration of the problem:

"Taking occasion thence," he writes, "I too began to reflect upon the earth's capacity for motion. And though the idea appeared absurd, yet I knew that others before me had been allowed freedom to imagine what circles they pleased in order to represent the phenomena of the heavenly bodies. I therefore deemed that it would readily be granted to me also to try whether, by assuming the earth to have a certain motion, representations more valid than those of others could be found for the revolution of the heavenly spheres.

"And so, having assumed those motions which I attribute to the earth farther on in the book, I found at length, by much long-continued application, that, if the motions of the remaining planets be referred to the revolution of the earth, and be calculated according to the period of each planet, then not only would the planetary phenomena follow as a consequence, but the order of succession and the dimensions of the planets, and of all the spheres, and the heaven itself, would be so bound together that in no part could anything be transposed without the disordering of the other parts and of the entire universe."

So Copernicus commends his book to the Pope, whose authority and scholarly fame will surely protect it from the bite of calumny. As for the detraction of those unskilled in mathematics, he will disregard it: *Mathemata mathematicis scribuntur* (Mathematics is written for mathematicians).

§ 3. THE NEW ASTRONOMY AND THE OLD PHYSICS

Copernicus follows the practice of Ptolemy and of other old writers on astronomy by beginning his treatise with a number of general physical propositions relating to the earth and to the universe as a whole. In the first four chapters of the *De revolutionibus* he keeps close to the corresponding introductory portions of the *Almagest*. Thus we read that the universe is spherical, for the form of the sphere is the most perfect and capacious, and so forth (I, 1; cf. *Alm.*, I, 2). The earth also is a sphere, if we neglect surface irregularities (I, 2). The reasons here adduced are the sound ones of the *Almagest*, I, 3: anyone traveling northward sees a proportionate increase in the elevation of the north pole of the heavens above the horizon; a given eclipse appears later in the day to dwellers in the East than to those in the West, and departing ships seem to sink gradually below the horizon. The ocean fills up the depressions in the earth's surface; land and water possess the same center of gravity, so as to cast upon the eclipsed moon an invariably circular shadow (I, 3; cf. Aristotle: *De caelo*, II, 14).

Chapter 4 brings Copernicus to a cardinal doctrine of classical astronomy: that the motions of the heavenly bodies are uniform, eternal, and circular or compounded of circular motions. This doctrine, which was probably of Pythagorean origin, was supported by the authority of Aristotle (*De caelo*, I, 2, 3, and II, 6), and Ptolemy gave it his formal assent (*Alm.*, III, 3), although Copernicus considered him to have departed from it in practice in admitting that the center of an epicycle might move non-uniformly about the center of its deferent (e.g., in *Alm.*, IX, 5). Copernicus regards circular motion of the planets as alone compatible with the regular recurrences which we observe in their phenomena. "For it is the circle alone which can bring back again what has already taken place" (I, 4). True, a planet does not appear to move uniformly, and its distance from us seems to vary. A heavenly body cannot move in a single circle at a variable rate, however, for this would argue variableness in the motive force, or else in the body moved, and the mind recoils alike from either explanation. Hence we must attribute the planetary inequalities either to the multiplicity of the component motions affecting the planets or to the earth's being displaced from the common center of those motions.

How, then, is the earth related, in respect of its position and possible motion, to the rest of the universe? Previous writers, says Copernicus, have generally assumed, almost as a foregone conclusion, that the earth is at rest at the center of the universe; but the matter is not beyond dispute. "For every apparent change of position is due, either to a motion of the object observed, or to a motion of the observer, or to unequal changes in the positions of both. . . . If, then, a certain motion be assigned to the earth, it will appear as a similar but oppositely directed motion affecting all things exterior to the earth, as if we were passing them by." (I, 5.) Now the daily rotation of the heavens is a motion affecting everything exterior to the earth, and "if you will allow that the heavens have no part in this motion, but that the earth turns from west to east, then, so far as pertains to the apparent rising and setting of the sun, moon, and stars, you will find, if you think carefully, that these things occur in this way" (I, 5). Next, concerning the position of the earth in the universe, Copernicus finds it almost universally held that the earth is at the center. But "if someone states that the earth does not occupy the center of the universe, but nevertheless does not admit that its displacement is so great as to be comparable with the sphere of fixed stars, though appreciable and obvious in comparison with the spheres of the sun and of the other planets;

if then he supposes that the motion of those planets will therefore appear non-uniform, being referred to a center other than
the center of the earth, he will perchance be able to offer a not
unfitting explanation of this non-uniform apparent motion"
(I, 5). That is, the planetary inequalities might reasonably be
explained by supposing the earth to be displaced from the center of the planetary motions by an amount comparable with
the distances of the planets, but incomparably less than the distance of the fixed stars. Such a displacement must indeed be
negligible compared with the dimensions of the universe, for
the horizon divides the zodiac and the celestial sphere into two
equal parts (I, 6; cf. *Alm.*, I, 5). It does not follow, however,
that the earth must be *at rest* at the center; rather it appears
incredible that a universe so immense should revolve in twenty-
four hours, while its least part, the earth, remains at rest.

Copernicus next addresses himself (I, 7) to the arguments
based on mechanical grounds which the ancients had directed
against all theories involving the motion of the earth or its displacement from the center of the universe. To understand the
objections of this type which Copernicus had to overcome, it is
necessary to revert for a moment to the cosmological ideas of
Aristotle (Chapter I, § 1, *supra*).

In the Aristotelian system the elementary bodies constituting
the earth and filling the whole region within the moon's sphere
differed from the ethereal bodies forming the surrounding heavens not only in their substance but also in their natural modes
of motion. Thus it was supposed that, whereas elementary bodies
moved naturally in straight lines, outward from the center of
the universe, or inward toward the center, celestial bodies revolved eternally in circles round the center. These celestial
motions were supposed to be maintained by virtue of an incorporeal "unmoved mover," or by a plurality of such movers, inspiring the spheres to an activity represented by their uniform
rotation. The rectilinear motions of the terrestrial elements, on

the other hand, were attributed to a sifting agency of space itself, whereby these elements were relegated by "natural motions" to their "natural places," i.e., to the layers in which they were respectively supposed to congregate. Such laws of motion were not derived from intelligent experimentation, but were suggested by mere appearances, support for them being sought in plausible deductions from very general statements about the supposed nature of things (many of them little more than popular maxims), or sometimes even from the etymology of the terms employed.

Thus, in this matter of natural motions, Aristotle lays it down (*De caelo*, I, 2) that every motion must be either rectilinear or circular, or compounded of the two, and that the most excellent motion is that which can go on unaltered forever. Now rectilinear motion cannot be indefinitely maintained in a finite universe without sooner or later being stopped at the boundary of that universe; it is hence inferior to circular motion, which can be so maintained. But the natural motion of each terrestrial element is manifestly rectilinear—fire and air move straight upward, earth and water straight downward. Hence there must be some superior element to which circular motion is natural, and this is readily identified with the ether composing the heavens and the heavenly bodies.

Now Copernicus so far belonged to his age as not to find any fault with mechanical principles of this sort, which indeed were not effectively challenged until about a century after his death. His concern was only to rebut Aristotle's and Ptolemy's application of such principles to prove that the earth must be at rest. For his own part, he employs closely similar, and to a modern mind equally artificial and worthless, mechanical arguments to prove that the earth is more probably in motion. We may summarize as follows the typical mechanical arguments with which Copernicus deals.

(*a*) A simple substance possesses a single natural motion,

directed toward, away from, or round, the center of the universe. Earth and water have rectilinear downward motions; air and fire have rectilinear upward motions. If the earth performed a daily rotation, this principle of simple natural motions would be violated. (Cf. Aristotle: *De caelo,* I, 2.)

(*b*) Heavy bodies tend to move in straight lines toward the center of the earth (as the center of the universe) and to come to rest there. They show no natural tendency to move in any other direction. Hence the whole earth, which is simply a collection of such heavy bodies, can have no natural tendency to move in any other direction; and no motion of the earth which was not natural could be eternal. (Cf. *De caelo,* II, 14; *Alm.,* I, 6.)

(*c*) If the earth were in motion, clouds and other bodies floating in the air would appear to be always traveling in the direction opposed to that motion. (Cf. *Alm.,* I, 6.)

(*d*) If the earth performed a daily rotation, the rapidity of its motion would need to be enormous, and anything rotating is more likely to throw bodies off than to draw them to itself. The earth would long since have been scattered abroad and life destroyed.

Let us look now at the way in which Copernicus meets these several objections (I, 8).

(*a*) He lays down the principle that every body, terrestrial as well as celestial, possesses a natural *circular* motion. So long as the body remains in its "natural place," this is the only motion that it does possess. Rectilinear motion, however, is superadded to this in any body which is out of its natural place. Thus rising and falling bodies appear to move perpendicularly upward or downward because, being parts of the earth, they partake of the earth's circular motion; when they are at rest on its surface, they possess this motion *only*.

(*b*) Copernicus does not explicitly distinguish this objection from (*a*), but it clearly vanishes if one is at liberty to add any (natural) motion of the earth to the apparent rectilinear motion of a terrestrial body. Moreover, in Chapter 9 Copernicus takes a view of gravity which has an important bearing on this point.

"I am of the opinion," he writes, "that gravity is nothing but a certain natural tendency to draw together, which is implanted in parts by the divine providence of the Maker of all things, that they may collect themselves into unity and completeness, being assembled into the form of a sphere. It is to be supposed that this influence resides also in the sun, moon, and other planets in order that, by its agency, they may remain in that globular shape in which they appear. Nevertheless, these bodies perform their divers circuits." (I, 9.)

Now if gravity be thus attributed to bodies demonstrably in motion, terrestrial gravity cannot be admitted as evidence that the earth is stationary.

(c) As for the behavior of clouds and other such bodies, we must suppose that a considerable proportion of the atmosphere, together with the bodies floating in it, is carried round with the earth in its motion, just as, in the traditional view, the upper portion of the atmosphere partakes of the diurnal motion of the heavens and carries round comets in its course.

(d) Copernicus denies that the earth would be disrupted by a diurnal rotation. For it is implied that such motion would be *natural,* producing effects contrary to those produced by *violence*. Why not rather fear the disruption of the universe, whose parts, in the traditional view, must move with a celerity incomparably greater than that involved in the rotation which it is now proposed to attribute to the earth?

Summing up, Copernicus concludes that it appears more probable that the earth is in motion than that it is at rest.

Can the earth, then, be regarded as a planet? If it possesses any motions, corresponding motions must appear in many external objects (I, 9), in accordance with the principle already laid down, of the reciprocity of apparent motions. Copernicus proceeds to apply this principle to the case of the sun's annual circuit.

"If [this circuit] be transposed from being a solar to being a terrestrial [phenomenon], and it be granted that the sun is at rest, then the risings and settings of the signs and the fixed stars, whereby they become morning and evening stars, will appear after the same manner [as before]. The stations, retrogressions, and progressions of the planets will also be seen to be a motion of the earth, not belonging to the planets themselves, but borrowed by them in their apparent behavior. Lastly, the sun himself will be deemed to occupy the center of the universe, all of which we are taught by the order in which the planets succeed one another, and by the harmony of the entire universe, if we will but look at the matter with both eyes, as they say." (I, 9.)

Thus in the first nine chapters of the *De revolutionibus* Copernicus clears away what his contemporaries would consider the chief objections to regarding the earth as one of the planets. He has next to establish the relation in which the planet earth stands to the other bodies of the universe.

§ 4. THE COPERNICAN UNIVERSE

Copernicus approaches his new geometrical scheme of the solar system through a discussion of the order in which the planets succeed one another with increasing distance from the center of the universe (I, 10; cf. *Alm.*, IX, 1). Ever since the time of Anaximenes (sixth century B.C.) the stars had been recognized, by common consent, as the most distant of the visible celestial bodies. They were thought to be situated upon, or within the substance of, a crystal sphere, which formed, in the Aristotelian system, the boundary of space. Its characteristic motion was the diurnal rotation about the earth in which all the seven planets participated. In order to account for Hipparchus' discovery of the precession of the equinoxes (Chapter

IV, § 2, *infra*), Ptolemy thought it necessary to attribute to the stellar sphere a slow rotation about the axis of the ecliptic, in addition to the diurnal one about the axis of the equator. Now, in the ancient systems of planetary spheres, each component sphere, as we have seen, possessed only a single rotation of its own. In order that the stellar sphere should conform to this rule, Ptolemy (in his *Planetary Hypotheses*) postulated a sphere exterior to that of the stars and possessing the diurnal rotation only; this motion it transmitted to the stellar sphere, the characteristic motion of which was now simply that required to account for the precession. During the Middle Ages, however, the idea established itself that the sphere of stars, besides showing a continuous rotation about the axis of the ecliptic, was subject to a small, periodic disturbance which caused the equinoctial points to oscillate about their mean positions (see Chapter IV, § 2, *infra*). In the time of Copernicus it was customary to attribute this oscillatory motion to the sphere of stars (which was called the *eighth* sphere, because it was additional to the systems of the seven planets), and to postulate two further spheres, called the ninth and the tenth spheres respectively, exterior to the eighth. Of these, the ninth imparted the precessional motion and the tenth (or *primum mobile*) transmitted the diurnal motion to all the systems lying within. Outside of all these spheres there was the motionless Empyrean—the abode of God and of the saints. This system is depicted in Fig. 6, which is reproduced from the 1539 edition of the *Cosmographicus liber* of Peter Apian, a contemporary of Copernicus.

Copernicus retains the traditional conception of a finite space bounded by the sphere of stars; but he places the sun instead of the earth at its center. The ancients supposed that the distances of the moon, the sun, Mars, Jupiter, and Saturn from the center of the universe increased in the order in which these bodies are here mentioned—presumably because the periods in which they complete their respective circuits of the zodiac in-

FIG. 6. The Universe According to Peter Apian (From P. Apian, *Cosmographia*, 1539. Courtesy of the British Museum)

crease in that order, from the moon's period of one month to Saturn's thirty years. As to the positions of Venus and Mercury in relation to the sun, however, there was much uncertainty. These planets, unlike Mars, Jupiter, and Saturn, always keep within moderate angular distances of the sun. If, as Ptolemy supposed, their paths lay between the earth and the sun, it was argued that they should occasionally be seen against the solar disc, and that, if they were dark bodies, they might even show phases, like the moon, when their illuminated hemispheres were partly turned away from us. If, on the other hand, they were more distant than the sun, as Plato taught in the *Timaeus,* there must be a great empty space between the sun and the moon, contrary to the prevailing belief that the sphere of one planet was contiguous to that of the succeeding one (see VI, § 2, *infra*).

Copernicus declares for the hypothesis of Heraclides, which he associates with the name of Martianus Capella (see § 6, *infra*), that Venus and Mercury describe circles with the sun as center; and he extends this hypothesis to the superior planets also. "If anyone takes occasion from this to refer Saturn, Jupiter, and Mars also to the same center, he will not err, provided he understands the size of their spheres to be such as to contain and surround the earth lying within" (I, 10). This assumption that Saturn, Jupiter, and Mars revolve about the sun in orbits which embrace the earth will explain why these planets are nearest to us when they rise in the evening and farthest from us when they set in the evening. "But, indeed, if all these planets depend upon one center, a space must be left between the outside of the sphere of Venus and the inside of the sphere of Mars, and an orb or sphere be separated off, concentric with the other spheres in respect of each of its boundaries, and capable of receiving the earth with her handmaiden the moon, and everything contained within the lunar sphere. . . . Wherefore we are not ashamed to confess that everything the moon goes round, with the center of the earth, travels through that great orb among the other planets, with an annual revolution round the sun. Further, that the center of the universe is in the sun, which remains unmoved, so that whatever motion it appears to possess should more truly be ascribed to the mobility of the earth. Also that the magnitude of the universe is such that, whereas this displacement of the earth from the sun is of appreciable magnitude compared with any of the other planetary spheres, it is

inappreciable in comparison with the sphere of the fixed stars."
(I, 10.) Copernicus thus arrives at a scheme of things best illustrated, in its broad outlines, by his well-known diagram (Fig. 7).

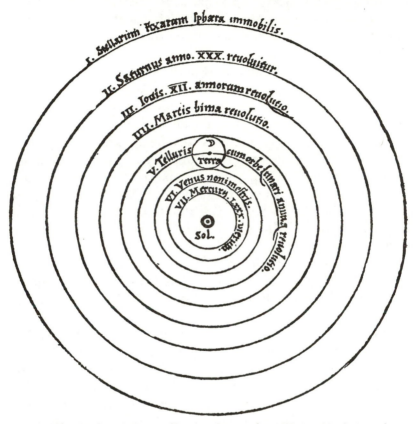

FIG. 7. The Universe According to Copernicus (From N. Copernicus,
De revolutionibus, 1543. Courtesy of the British Museum)

"In the midst of all dwells the sun. For who could set this
luminary in another or better place in this most glorious temple, than whence he could at one and the same time lighten the
whole? ... And so, as if seated upon a royal throne, the sun rules

the family of the planets as they circle round him. . . . That nothing of these things appears in the fixed stars proves their immense distance above us, which is sufficient to cause even the annual orbit, or its appearance [in the stars], to vanish from our eyes." (I, 10.)

This last sentence calls for some further explanation. It is a matter of common experience that an observer in motion (e.g., a passenger in a moving railway train) is aware of progressive changes in the apparent directions and grouping of stationary objects in his vicinity. Such apparent displacement of an object owing to a change in the position of the observer is known as *parallax*. It had been recognized from antiquity that an annual revolution of the earth about the sun should cause each star to appear to revolve about its mean position with an annual periodicity. Thus Aristotle argued that, if the earth ever moved away from the center of the universe, "it is necessary that there should be transitions and mutations of the fixed stars (*De caelo*, II, 14). For many centuries astronomers sought in vain for any traces of such parallax; and their failure constituted one of the most serious objections to all planetary hypotheses involving a translatory motion of the earth. Aristarchus of Samos and Copernicus were both forced to meet this objection to the heliocentric system by postulating that the distance of the stars from us must be incomparably greater than the diameter of the earth's orbit, so that our annual motion could have no sensible effect upon their apparent positions. The annual stellar parallax was first detected, and measured with reasonable accuracy, a little more than a century ago (see Chapter VIII, § 6, *infra*).

§ 5. THE STATUS OF THE COPERNICAN THEORY

In the discussion of the philosophical first principles of astronomy which was carried on intermittently from the time of Plato to the sixteenth century, a fundamental distinction was drawn between two types of planetary hypotheses (cf. Chapter I, § 2, *supra*). Both of these aimed at "saving the phenomena" (accurately representing the facts of observation) by postulating for each planet a set of simple and regularly recurrent motions which, when compounded together, should produce the complicated motion exhibited by the body in question. But, on the one hand, there were hypotheses which, besides being adjusted to fit the facts of observation, were further restricted to conform to those principles of celestial physics to which the planets, as physical bodies, were thought to be subject. On the other hand, there was a more restricted type of hypothesis the sole justification of which was that it conformed to, and systematized, the motions of a planet observed in the past and that it enabled the future motions to be successfully predicted. It made no attempt to explain, by reference to physical principles, *why* the planet should possess the component motions attributed to it; and if two or more alternative sets of such motions represented the planet's behavior equally well, it afforded no criterion (apart from mere convenience) for preferring any particular one of them. A typical hypothesis of the former type was the planetary system of Aristotle; the system of Ptolemy, with its accumulated empirical refinements, was a complex of hypotheses of the latter type.

It was natural for the earliest readers of the *De revolutionibus* to ask what relation Copernicus supposed his theory to have to the physical facts. Did he believe that the earth actually possessed the motions which he assigned to it? Or did he regard his theory as a mere computing device, intended to facilitate the construction of improved planetary tables? An unpredictable complication of the problem was introduced by the presence of Osiander's Preface (see Chapter II, § 4, *supra*), the tone of which seemed at variance with that of the rest of the book. That Copernicus did not regard his theory as a mere computing device is suggested by the following considerations:

(*a*) He designates his doctrine of the earth's mobility by such words as *opinio, doctrina, cogitatio.*

(*b*) He refers to himself as the man "into whose mind it came to dare to imagine some motion of the earth, contrary to the received opinion of the mathematicians and well-nigh contrary to common sense" (Dedicatory Preface). He would scarcely have written in such terms about a computing device, since such an artifice is exempt from all physical restrictions.

(*c*) Foreseeing that some may condemn his opinion as unscriptural, Copernicus likens such an attitude to that of Lactantius, a Latin Father of the Church, who poured scorn on the idea that the earth was a sphere (*Divinae institutiones,* III, 24). "It ought not to surprise scholars," writes Copernicus, "if they pour such scorn on us also" (Dedicatory Preface). Copernicus here implies that his doctrine is comparable to that of the sphericity of the earth; and this point is made at the risk of offending the Pope by the reference to the ignorance of Lactantius. We may further recall the space devoted and the important place assigned in the *De revolutionibus* to proofs of the physical admissibility of the earth's motion.

(*d*) On the other hand, in referring to the detailed geometrical schemes which represent the terrestrial and planetary motions, he ordinarily uses the word *hypothesis* (e.g., I, 11; IV, 3; V, 4).

(*e*) When, on several occasions, he describes or suggests alternative eccentric and epicyclic systems, his choice between them is in each case governed by reference to tradition or convenience (e.g., III, 15; III, 20; IV, 3; V, 4): "which of the two [systems] exists in the heavens," he writes in one such case, "it is not easy to distinguish" (III, 15).

It appears likely, then, that Copernicus intended his doctrine of the earth's motion to be understood as a statement of physical fact. It is possible, however, that the matter may not have presented itself to Copernicus in the light of a straight choice between statement of physical fact and mere computing device. Exposed in Italy to the ideas of the Platonic revival, with their strong Pythagorean tincture, he would become familiar with a subtle, and characteristically modern, conception of the "truth" of a scientific explanation. It was the ideal of that movement to substitute for the purely qualitative, classificatory science of the Scholastics the mathematical analysis of natural phenomena, and to assign the highest degree of truth to that theory which most simply and symmetrically accounted for the observed facts and established the necessity of the quantitative relations observed to connect them. In the spirit of this movement, Copernicus would be disposed to embrace that theory from which followed, by mere mathematical necessity, the greatest number of fundamental celestial phenomena. By transferring his origin of reference from the earth to the sun, Copernicus succeeded in reducing by more than half the number of arbitrary circular motions which Ptolemy had been obliged to postulate. At the same time he was able to demonstrate the necessary connection between his assumptions and certain characteristic planetary vicissitudes (for example, the periodically recurrent retrogressions of the planets, and the fact that Mars appears brightest when in opposition to the sun) which would otherwise have had to be regarded as mere unrelated coincidences. The tone of the passages quoted above in the two preceding sections rather suggests that Copernicus applied such a mathematico-aesthetic criterion to the rival planetary theories, whereas his opponents were arguing in terms of verifiable physical facts. It was to the judgment of the mathematically trained that Copernicus appealed in his Dedicatory Preface; and it was from their ranks that his earliest adherents were drawn.

§ 6. PRECURSORS OF COPERNICUS

In reviewing the contents of the *De revolutionibus,* we shall have occasion to make numerous references to the *Almagest* of Ptolemy. From its pages Copernicus derived many of his observational data and geometrical devices, with or without acknowledgment. The ideas which constitute Copernicus' essential contribution to astronomy, however, were certainly not derived from Ptolemy; if we are to trace them to any source, it must be sought in the speculations of a few classical and medieval thinkers who stood apart from the main current of opinion. Certain of the writings or recorded teachings of these men were known to Copernicus; how far they anticipated his doctrines may best be gathered from a study of several critical passages—in particular of those to which explicit reference is made in the *De revolutionibus.*

We begin, then, with the extract from the *De placitis philosophorum* of pseudo-Plutarch, quoted in Greek in the Dedicatory Preface to Paul III:

"It is commonly maintained that the earth is at rest. But Philolaus the Pythagorean held that it revolves round the Fire in an oblique circle in like manner to the sun and moon. Heraclides of Pontus and Ecphantus the Pythagorean suppose the earth to move, not with a motion of translation, but after the manner of a wheel turning upon an axle about its own center, from west to east." (*De plac. philosoph.,* III, 13.)

In the same work, and only a few pages before the foregoing passage, we read:

"Aristarchus places the sun among the fixed stars, and holds that the earth moves round the sun's circle" (*ibid.,* II, 24). Copernicus does not quote this passage; he mentions Aristarchus only three times (III, 2, 6, and 13) and in no case refers to his heliocentric theory. There is, however, in the original manuscript (at the close of I, 11) a long passage which Copernicus has scored out but which contains the words: "Although we admit that the motion of the sun and moon might also be demonstrated on the assumption that the earth is immovable, this agrees less with the other planets. It is probable that it was on these and similar grounds that Philolaus judged that the earth moves, and some say that Aristarchus of Samos was of the same opinion." *(De rev.,* 1873 ed., p. 34.)

In the Dedicatory Preface and again in I, 5, Copernicus alludes to Cicero's account of Hicetas (see Chapter I, § 1, *supra*); he evidently has the following passage in mind:

"Hicetas of Syracuse, as Theophrastus relates, is of the opinion that the heavens, the sun, the moon, the stars, and, in short, all the heavenly bodies are at rest, and that nothing in the universe moves except the earth; and as the earth turns and twists itself with extreme rapidity about its axis, all the same appearances are produced as if the heavens were in motion and the earth stood still" (Cicero: *Academica,* II, 39).

Lastly, in constructing his scheme of the solar system (I, 10), Copernicus refers to the account of the "Egyptian" (restricted heliocentric) system (see Chapter I, § 1, *supra*) given by Martianus Capella:

"For although Venus and Mercury show daily risings and settings, yet their orbits by no means surround the earth, but encircle the sun with a wider circuit. In short, they make the sun the center of their orbits" *(De nuptiis Philologiae et Mercurii,* Liber VIII).

Copernicus appears to acknowledge no inspiration from later writers than Martianus, but some of his ideas seem to have been anticipated by Nicolaus de Cusa, or Cusanus, a fifteenth-century ecclesiastic. There are passages in his *De docta ignorantia* (written about 1440 and first printed in 1514) which are entirely in the spirit of Book I of the *De revolutionibus.* Thus, having laid it down that the universe is infinite, and can therefore have no center, Cusanus continues:

"Since, then, the earth cannot be the center [of the universe], it cannot be entirely devoid of motion. . . . And since we cannot perceive motion except by reference to something fixed, such as poles or centers which we presuppose in measuring motions . . . we find ourselves to be at fault in all things, and we are surprised when we do not find the stars agreeing in their positions with the rules of the ancients, because we imagine that these held right views about centers and poles and measurements." (*Op. cit.*, II, 11.) Again:

"Now it is clear to us that the earth is really in motion, though this may not be apparent to us, since we do not perceive motion except by a certain comparison with something fixed" (*ibid.*, II, 12).

To sum up, it seems clear from such passages as these:

(*a*) that the possibility of explaining the apparent diurnal rotation of the celestial sphere by assuming an equal and opposite rotation of the earth was fairly widely recognized in antiquity:

(*b*) that certain of the ancients had a theory that Venus and Mercury describe circles about the sun in the center;

(*c*) that certain Pythagoreans suggested that the earth might possess a motion of translation, although their system was in no sense a heliocentric one;

(*d*) that Aristarchus anticipated the Copernican system in an undeveloped hypothesis;

(*e*) that Copernicus was acquainted with all these surmises when he developed his own system.

It appears, then, that the reformative ideas which we associate with Copernicus are not to be regarded as original products of his genius. His great contribution to astronomy lay rather in his development of those ideas into a systematic planetary theory, capable of furnishing tables of an accuracy not before attained, and embodying a principle the adoption of which was to make possible the triumphs of Kepler and Newton in the following century.

§ 7. A GENERAL SURVEY OF THE COPERNICAN SYSTEM

We have in this chapter summarized the underlying physical principles and sketched the broad outlines of the new planetary system in such general terms as Copernicus himself employs. We have next to examine in some detail the geometrical schemes by which he seeks to represent the principal celestial motions. The following three chapters (IV–VI), which embody the essential substance of the *De revolutionibus* (Books III–VI), are necessarily those which make the greatest demands upon the student. Accordingly it may be appropriate to include at this point a brief, nontechnical summary of these chapters for the benefit of the general reader who does not wish to become involved in the geometry of an old-time planetary theory.

Copernicus begins by explaining how certain well-known celestial phenomena can be readily accounted for by reference to corresponding motions which he proposes to attribute to the earth (I, 11, and III). Thus, as we have seen, the daily turning of the celestial sphere, which causes the rising and setting of heavenly bodies, is to be regarded as merely an appearance produced by a diurnal rotation of the earth upon its polar axis. The seasonal travel of the sun through the constellations, and the fluctuations in the length of the day, can, in like manner, be explained as consequences of an annual revolution of the earth round the sun in a plane obliquely inclined to the polar axis. Moreover, by supposing that axis to describe very slowly a cone in space, Copernicus was able to explain why the pole of the heavens appears to trace out a circle among the stars in a period which must amount to about 26,000 years. The ancients had been aware of this phenomenon and had attributed it to a motion of the celestial sphere itself, assuming the earth's axis to remain fixed; but the Copernican view of the matter prepared the way for the dynamical interpretation which Newton was later to provide. Copernicus was led by his own and other men's observations to believe that the pole traced a complicated path through the heavens; and much of Book III is occupied with building up a theory of its motion which has now lost all but historic interest. He represented the circulation of the earth round the sun essentially by interchanging the roles of earth and sun in Hipparchus' theory (shown in Fig. 2, *supra*); but he introduced additional complications in order to allow for supposed periodic fluctuations in the amount of eccentricity of the earth's orbit and in the rate at which its apse-line swings round from west to east (as al-Battani had discovered and as Copernicus confirmed).

The contributions of Copernicus to lunar theory, contained in Book IV of the *De revolutionibus* (and summarized in Chapter V, *infra*), are of minor significance from the cosmological point of view. However, by means of a skillfully chosen combination of epicycles, the relative sizes of which were determined from selected observations, he was able to give a fair representation of the moon's apparent motion from night to night without assuming (as Ptolemy had done) variations in the moon's distance from the earth, and hence in its apparent size, out of all proportion to what observation reveals. Copernicus repeated the calculations by which Ptolemy had sought to ascertain the distances of the sun and the moon from the earth, but without arriving at significantly different estimates of these quantities. The ancients had a fairly accurate idea of the moon's distance (about sixty times the radius of the earth); but their estimate of the sun's distance (about 1200 times the earth's radius) was little more than one-twentieth of its true value, and no radical improvement was made upon it until toward the close of the seventeenth century.

The core of the *De revolutionibus* is represented by Book V (analyzed in Chapter VI, *infra*), in which the Copernican theory of the planetary motions is developed. Copernicus begins by showing that, according to his hypothesis, the characteristic apparent motions of the planets, with their stationary points and arcs of retrogression (Chapter I, § 1, *supra*), can be simply accounted for, once for all, as optical effects resulting from the orbital revolution of the terrestrial observer round the sun. This explanation constituted the chief claim of the Copernican theory to acceptance, for it eliminated the large epicycles which Ptolemy and his successors had been compelled to introduce arbitrarily into each planet's economy in order to account for the characteristic planetary motion. The remaining fluctuations in the rates at which the planets move are, however, still represented in the Copernican scheme by means of such combinations of circular motions as Ptolemy had employed for that purpose, the relative sizes of the circles being determined from suitably chosen sets of observations. The necessary calculations are first performed with observations cited by Ptolemy and then repeated, as a check, with others made or collected by Copernicus himself.

Copernicus found that it was possible, by observing a planet from two points on the earth's orbit, to determine the relative distances of the planet and the earth from the sun, i.e., to specify the radius of the planet's orbit in terms of the radius of the earth's orbit as unit. The results thus obtained show remarkable agreement with modern estimates of the "mean distances" of the planets from the central luminary.

The planetary theory of Copernicus culminated in the construction of tables for computing a planet's position in the sky at any given time. The substitution of the heliocentric for the geocentric hypothesis simplified the construction of such tables; otherwise it brought little direct or immediate advantage to the practical astronomer. Only gradually and indirectly did the Copernican revolution, by opening the way to a prodigious advance in dynamical astronomy, contribute to the precision of modern planetary and lunar tables.

After this hasty glance through some of the more technical pages of the *De revolutionibus*, we must now attempt to expound their contents in greater detail.

4

The Copernican System: Theory of the Earth's Motion

HAVING ASSIGNED TO THE EARTH A PLACE AMONG THE PLANETS, Copernicus has next to specify what motions must be attributed to it in order to account for the diurnal and seasonal phenomena and for certain slow, secular changes in the earth's axis of rotation and in the elements of its orbit.

§ 1. DIURNAL AND ANNUAL MOTIONS OF THE EARTH

The motions which Copernicus attributes to the earth (I, 11) are essentially as follows (we shall consider the refinements of the theory in the succeeding two sections of this chapter):

(*a*) a diurnal rotation from west to east about the polar axis;

(*b*) an annual revolution about the sun from west to east in the plane of the ecliptic;

(*c*) a variation in the inclination of the earth's axis to the line joining sun and earth, of nearly the same period as (*b*), but of opposite tendency.

The motion (*a*) accounts for the daily apparent rotation of the celestial sphere about the poles of the heavens. The motion (*b*) is the cause of the sun's yearly apparent circuit through the zodiacal constellations. The motion (*c*), introduced as part of the explanation of the phenomena of the seasons, requires a little further explanation.

It was obvious to Copernicus that seasonal changes result from variations in the attitude of the earth's axis to the sun; and he saw that it would be possible to account for such variations by supposing the earth's axis to preserve an approximately invariable direction relative to the fixed stars. For let ABCD be the earth's circular orbit in the plane of the ecliptic, with the sun at its center E (Fig. 8). Let EA, EB, EC, ED be directed

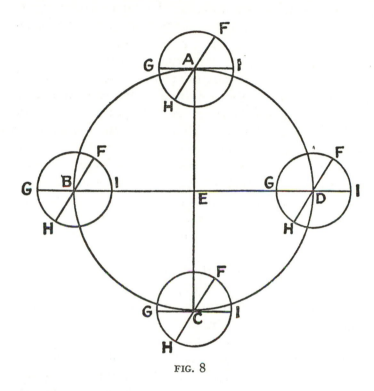

FIG. 8

toward the first points of the zodiacal signs of Cancer, Libra, Capricornus, and Aries respectively. Let FGHI be the earth's equator, intersecting the plane of the ecliptic in the diameter GAI. Draw FAH perpendicular to GAI in the plane of the equator, F being to the south and H to the north of the ecliptic. Let the earth now travel round its orbit while GI and FH retain unaltered directions in relation to the fixed stars. When the earth is at A, it will be the winter solstice, and the sun will appear south of the equator, at the first point of Capricornus. If now the earth travels to B, the sun will appear to lie on the equator in the direction GI, i.e., at the first point of Aries, and it will be the spring equinox on the earth. When the earth reaches C, the sun will appear at the summer solstitial point; and when it reaches D, the sun will once again be on the equator, and we shall have the autumn equinox.

We have here the explanation of the seasons which is given in modern textbooks; the condition for its validity is that the earth's axis of rotation should, to all appearance, continue to point in the same direction, while the earth's center describes a circle about the sun. The mechanical ideas of a later age conduced to the instinctive expectation of such behavior in an unconstrained rotating body. Copernicus, however, adopted the attitude of the classical writers on astronomy, who regarded a planet, epicycle, etc. as being carried round the center of its revolution like an object attached to the rim of a wheel (cf. Chapter I, § 1, *supra*). Such an object must present always the same aspect to the hub of the wheel, unless affected by extra motions of its own. Thus, for example, the ancients regarded the moon as having no rotation of its own *because* it presents always the same face to the earth (Aristotle, *De caelo*, II, 8). And similarly Copernicus must have thought that the motion (*b*), of itself, would cause the earth's axis of rotation to remain inclined at a constant angle to the line joining earth and sun, thus describing a cone in space, and admitting of no seasonal fluctuations in the angular distance of the sun from the equator. He therefore introduced the motion (*c*), the function of which was to keep the direction of the earth's axis invariable throughout the year, by causing it to describe a cone in space which should just neutralize the conical motion supposed to be produced by the revolution (*b*). Kepler, however, writing about fifty years after Copernicus, recognized the superfluity of this motion (c): "The said motion," he wrote, "is, in truth, not motion at all, but rather rest" (see *Mysterium cosmographicum,* cap. I, Author's Notes g and p; Frisch's edition, I, 119 and 121). This principle, thus admitted by Kepler in 1596, received more explicit formulation from Newton a century later: "Suppose an uniform and exactly sphaerical globe [rotating about an axis]. . . . Because this globe is perfectly indifferent to all the axes that pass through its centre, nor has a greater propensity to one axis or to one situation of the axis than to any other, it is manifest that by its own force it will never change its axis, or the inclination of it." (*Principia,* Book I, Prop. LXVI, Cor. 22; Motte's translation.)

Besides the periodic diurnal and seasonal phenomena, there are certain slow changes in the relation of the earth to the sphere of stars which constitutes the ultimate system of reference; and Copernicus makes use of the motion (c) in his explanation of these changes. His treatment of this problem occupies the first twelve chapters of Book III of the *De revolutionibus,* to which we shall now turn.

§ 2. THE PRECESSION OF THE EQUINOXES

The earth's motions of diurnal rotation, and of annual revolution round the sun (§ 1, *supra*), define two great circles on the celestial sphere: (a) the *celestial equator,* in which the plane of the earth's equator cuts the celestial sphere, and which has as its poles the two points about which the sphere appears daily to rotate, and (b) the *ecliptic,* in which the plane of the earth's annual orbit cuts the celestial sphere. If the earth's axis of rotation retained a constant direction in relation to the sphere of stars, and if the plane of its orbit were likewise invariable, then the two *equinoctial points,* in which these two great circles intersect, would remain fixed in relation to the stars. In actual fact, however, the equinoctial points are not stationary, but move slowly round the ecliptic, relative to the stars, in the opposite direction to that in which the sun performs its yearly circuit. In consequence, a progressive increase is observed in the *longitudes* (Chapter I, § 1, *supra*) of all the stars, which are measured eastward from the vernal equinoctial point (where the sun crosses the equator at the spring equinox). This steady increase in the stellar longitudes was recognized, in the second century B.C., by Hipparchus of Rhodes (*Alm.*, VII, 2), and Ptolemy interpreted it as due to a slow eastward rotation of the sphere of the stars about the poles of the ecliptic. In consequence of this motion, the period in which the sun completes a circuit of the heavens relative to the stars—the *sidereal year*—exceeds (by about 20 minutes) the period in which it completes a revolution relative to the equinoctial points—the *tropical year*. The consequent recurrence of the spring equinox and the beginning of a fresh seasonal cycle *before* the expiry of a complete sidereal year from the last equinox have caused the whole phenomenon to be termed the *precession of the equinoxes* (Copernicus uses the terms *praecessio, anticipatio, praeventio*).

Copernicus begins his third book with a brief historical introduction to the problem of precession (III, 1). He then proceeds to explain the phenomenon as due not to a motion of the stars but to a continuous alteration in the plane of the earth's equator, whereby the earth's axis of rotation describes a cone about the axis of the ecliptic. In order to account for such a circulation of the earth's axis, Copernicus merely postulates a slight inequality between the periods in which the two motions designated as (*b*) and (*c*) in the last section, are respectively completed. That is, he makes the period of the variation in the inclination of the earth's axis to the line joining earth and sun slightly less than the period of revolution of the earth round the sun, the *obliquity* (inclination of the equator to the ecliptic) meanwhile retaining a fairly constant value. "For . . . the two revolutions, I mean the annual one affecting the inclination, and that of the earth's center, are not exactly equal, the restoration of the inclination to its original value occurring a little before the center completes its circuit. Whence it necessarily follows that the equinoxes and solstices appear to fall early, not because the sphere of fixed stars moves eastward, but rather because the equator moves westward, remaining inclined to the ecliptic according to the measure of the inclination of the axis of the terrestrial globe." (III, 1.) This conception of precession as due to a motion of the earth's axis eventually established itself, especially after it had enabled Newton to give a dynamical explanation of the whole phenomenon, although nowadays we have further to take account of minute alterations in the plane of the *ecliptic,* unsuspected by Copernicus.

The diversity in the estimates of the rate of precession (i.e., of the rate of regression of the equinoctial points on the ecliptic) made throughout the centuries had given rise to the idea that this rate was subject to considerable variations, and even that the equinoctial points, instead of moving continuously round the ecliptic, relative to the stars, oscillated about their mean positions, with a period and amplitude which were variously estimated. This idea of a *trepidation* of the equinoctial points seems to have originated during the Alexandrian period; it prevailed among Indian writers on astronomy from about the fourth century A.D.; from them it spread to the Arabs, although the great al-Battani remained unconvinced of its truth. In the Latin version of the *Alfonsine Tables,* which became known to astronomers in northern Europe early in the fourteenth century, it was assumed that the sphere of stars was subject to a continuous precessional motion *plus* an oscillation about its mean position.

Copernicus, while rejecting the idea of the "trepidation" of the equinoctial, seeks to establish that the rate of precession is non-uniform, and to ascertain its periodicity and mean value, by reference to a series of recorded determinations of the longitudes of the bright stars Spica, Regulus, and β Scorpii, made by the ancient astronomers Timocharis (third century B.C.), Hipparchus, Menelaus, and Ptolemy, by the Arab al-Battani, and by Copernicus himself (III, 2). Upon these data he bases a geometrical theory intended to account, both for the inequality in the rate of precession, and for a slow diminution in the values of the obliquity of the ecliptic recorded through the ages. Unfortunately, Copernicus never made any allowances for the possibility of serious errors in the actual observations of his predecessors (even supposing these to be genuine), or of textual corruptions in the manuscripts by which the results had been handed down. As Delambre remarks: "Après avoir renversé le système des anciens, il n'ose même suspecter leurs observations." This uncritical acceptance of recorded observations led Copernicus, in this instance, to construct a laborious theory of precession, having but little relation to the facts as they are at present understood, although perhaps of sufficient interest to justify a brief study.

Copernicus believed that the fluctuations in the rate of precession and in the value of the obliquity were related phenomena. He sought to account for them by attributing two independent oscillatory motions to the earth's polar axis (III, 3). Each of these oscillations is to be performed "after the manner of suspended bodies," the motion taking place "over the same course, between two limits, most rapidly in the midst . . . and slowest at the extremities." (They are, in fact, what we should now call simple harmonic motions, as is clear from III, 4). The motions are applied to the polar axis in the following manner:

Let ABCD be the ecliptic (Fig. 9), E its north pole, A and C the solstices, and B and D the equinoxes. Draw the solstitial colure AEC, and let F and G be the positions of maximum and minimum displacement of the pole of the equator from E, respectively corresponding to the maximum and the minimum

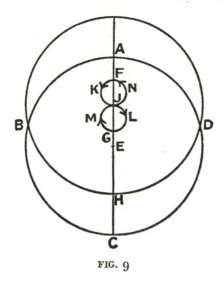

FIG. 9

obliquity of the ecliptic. Let BHD be the mean equator (cutting the ecliptic in the mean equinoctials B, D), and let J be its pole. Let J and the mean equator turn slowly and uniformly about E from east to west (clockwise in the figure). This uniform motion constitutes the mean precession and is accounted for, as we have seen, by a slight disparity between the rate of the annual motion of the earth and the rate of the compensatory motion of the earth's axis.

Now suppose the pole of the equator to be affected by two simultaneous oscillations of the kind described above:

(*a*) an oscillation between F and G;

(*b*) an oscillation perpendicular to FG, completed in half the period of (*a*). These two motions are to account for changes in the obliquity and in the rate of precession, respectively, and their joint effect is to cause the pole to describe the figure eight FKJLGMJNF on the celestial sphere (the opposite pole, of course, meanwhile traces out a corresponding figure in the reverse sense).

Copernicus has next to satisfy the demands of traditional physics by showing that rectilinear oscillations such as those described above can be produced by compounding certain uniform circular motions; he proceeds as follows (III, 4):

Let the straight line AB (Fig. 10) be divided into four equal

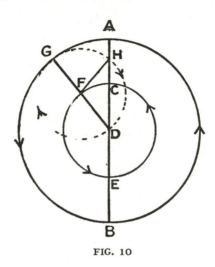

FIG. 10

parts at C, D, E. With center D and radii DA, DC, draw two circles. Let F be any point on the inner circle, and, with center F and radius FD, draw a circle cutting AB at H, D. Let F describe the circle CFE, and let H describe the circle GHD with twice the velocity of F in its circle, and in the opposite direction. Then

$$\angle GFH = 2 \cdot \angle GDH,$$

and it easily follows that H must lie constantly on the straight line AB and must oscillate between the limits A and B. (Actually $DH = DG \cdot \cos \angle GDH$, and the motion is simple harmonic.) The next step is to show that, in consequence of the oscillations of the pole of the equator, the equinoctial points vary about their mean positions and the obliquity fluctuates about its mean value, with oscillations of the same kind as those postulated for the pole (III, 5). To show that this is true of the oscillations of the equinoctials, let the mean equator AEB cut the ecliptic CED at the mean equinoctial point E (Fig. 11), and

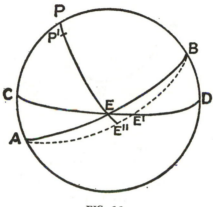

FIG. 11

let the pole P of the equator suffer a small displacement to P' at right angles to the solstitial colure APB. Then the equator is displaced into the position shown by the dotted line AE"B (where $P'E'' = PE = 90°$), cutting the ecliptic at the new equinoctial point E', where EE' is approximately PP' multiplied by a constant (the cosecant of the obliquity). Hence as P' oscillates about P, the equinoctial E' must oscillate about its mean position, E, in a similar manner; this oscillation is, of course, superimposed upon the steady retrograde motion of the equinoctial accounted for in III, 3, and it operates alternately to increase and to decrease the rate of precession, which is thus subject to a periodically recurring cycle of changes.

Copernicus typifies this cycle by means of the circle ABCD

(Fig. 12), the rate being least at A, increasing through its mean value at B, being greatest at C, and decreasing through its mean value at D. When any quantity shows periodic changes

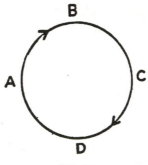

FIG. 12

regulated according to such a cycle, it should be possible to recognize the occasions of its passing through the stages represented by the points A, B, C, D, respectively. Copernicus accordingly seeks to deduce, from an examination of the extant observations, the dates at which the rate of precession passed through these four points in its cycle and hence the total duration of the cycle (III, 6). He finds this period to be about 1717 years, the minimum value of the rate of precession appearing to fall roughly midway between the dates of Timocharis and Ptolemy. This is, accordingly, the period in which, in Copernicus' scheme, the pole performs a complete transverse oscillation about its mean position, the period of oscillation assigned to the obliquity being twice as long (3434 years).

The mean rate of precession is deduced from the total precession in 1717 years (for which period the inequality cancels out), and it is found to be 1°23'40" per century (modern value 1°23'46"), the equinox performing a complete circuit of the ecliptic in 25,809 years.

Copernicus' mode of representing periodic variations such as those affecting the rate of precession, and of determining the values of quantities subject to them, must now be explained; it was essentially as follows:

Consider a circle ADBC (Fig. 13), with center O and two perpendicular diameters, AB and CD. Take any point P on the circle, and draw PM perpendicular to AB. Now let P move round the circle so that the ∠COP increases uniformly with the time. Then the oscillation of M about O represents the type

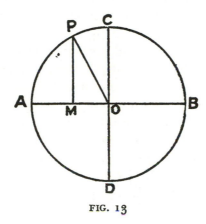

FIG. 13

of fluctuation of a point about its mean position, or of a quantity about its mean value, now being considered. The ∠COP, which increases steadily with the time, and which governs the value of the variable quantity in the manner indicated, represents what Copernicus calls the *argument of the anomaly*, or, more shortly, the *anomaly*. (For example, the anomaly of the precession, in Copernicus' scheme just described, increases through 360° in 1717 years.)

Copernicus tabulates the mean precessional motion and the increase in its anomaly for daily and yearly intervals (III, 6). A knowledge of the anomaly is, in fact, necessary in the calculation, for any given date, of the *equation of the equinoxes* (the angular distance between the true and the mean equinoctials). Before it can be so employed, however, the maximum possible value of this equation must be determined. This problem corresponds to that of finding the amplitude of a simple harmonic motion when the phase and the displacement from the mean position are given; Copernicus proceeds somewhat as follows (III, 7): He finds that in the 432 years that elapsed between two measurements of stellar longitudes, made by Timocharis and by Ptolemy respectively, the equinoctial point suffered an apparent displacement of $4°$ $20'$, which differs from the *mean* precessional motion for 432 years (viz., $6°$ $0'$) by $1°$ $40'$. His data further led him to assume that, roughly midway between the observations of Timocharis and Ptolemy, the rate of precession must have passed through a minimum, when the mean and apparent equinoctials would coincide and the argument of the anomaly would be zero. Hence the anomaly must have had equal and opposite values at the beginning and end of the period of 432 years; and since, during this interval, the anomaly must have increased through $(360 \times \frac{432}{1717})°$ or $90°$ $35'$, its initial and final values may be represented in a diagram (Fig. 14) constructed on the principle of Fig. 13 by the points P_1, P_2 where $\angle P_1OC = \angle COP_2 = \frac{1}{2}(90°\ 35')$

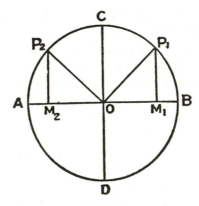

FIG. 14

$= 45° 17\frac{1}{2}'$. Draw P_1M_1, P_2M_2 perpendicular to AB. In this figure, AOB may be regarded as an arc of the ecliptic (treated as a straight line), having the mean equinoctial O at its center, and M_1, M_2 as the true equinoctials at the beginning and end of the 432 years. Then the arc M_1M_2 represents the difference between the mean and the apparent precessional motions of that period, which (as already stated) was 1° 40'. Hence

$$OM_1 = OM_2 = 50', \text{ and}$$

$$OA = \left(50 \times \frac{\text{radius of circle}}{\text{half-chord of } 90° \, 35'}\right)'$$

$= (50 \times \frac{10000}{7107})'$ (from the Table of Chords)

$= 1° 10'$, which is the required maximum departure of the true from the mean equinoctial.

The equation of the equinoxes can now be computed for any given date (III, 8): given the amplitude and phase of a simple harmonic motion, it is possible to find the corresponding displacement of the moving point. Copernicus tabulates the equation for each 3° of the anomaly from 0° to 360°; and he provides *radices,* specifying the place of the mean equinox and the value of the anomaly, at certain standard epochs. These radices, in conjunction with the tables, can be used to determine the place of the apparent equinox for any given date (III, 11, 12).

Considering next the obliquity of the ecliptic, Copernicus shows that the recorded estimates of the obliquity are consonant with the theory that this quantity oscillates between the values 23° 28' and 23° 52' in a complete period of 3434 years (III, 10). The above-mentioned table of the equation of the equinoxes includes a column of corrections to be added to 23° 28' to obtain the true obliquity corresponding to each 3° of anomaly.

The erratic motion of the vernal equinoctial point upon the ecliptic makes it, in Copernicus' opinion, no fit origin from which to measure the longitudes of the fixed stars. He therefore reckons his longitudes from the bright star "at the head of Aries" (γ Arietis); but his star catalogue (II, 14) is otherwise a mere *réchauffé* of Ptolemy's catalogue (*Alm.*, Books VII and VIII). The star catalogue in the *Almagest* is the earliest extant synopsis of its kind. It has long been disputed whether Ptolemy himself made the observations upon which it is based or whether he merely brought up to date the catalogue known to have been constructed nearly three centuries earlier by Hipparchus, which has not survived. The view that Ptolemy himself independently determined at least a substantial proportion of his star places has gained ground following Vogt's partial restoration of Hipparchus' original catalogue from data in his surviving *Commentary on the Phenomena of Eudoxus and Aratus* (*Astr. Nachr.*, 1925, Nos. 5354–55).

§3. THE EARTH'S ECCENTRIC

The theory of precession leads Copernicus, by a natural transition, to the problem of giving an exact representation of the earth's annual revolution round the sun. For if the equinoctial points traverse the ecliptic at a variable rate, it must follow that successive tropical years cannot be all of the same duration. And when evidence of this irregularity was sought among determinations of the length of the tropical year made throughout the ages, the question arose whether the discrepancies brought to light might not be due in part to certain inequalities affecting the *sun's* apparent motion relative to the stars.

Ptolemy recognized but one solar inequality, viz., the annual cycle of changes in the sun's rate of motion in longitude whereby equal arcs of the ecliptic are, in general, described in unequal times (*Alm.*, III, 4); we have seen how Hipparchus represented this inequality by assuming the sun uniformly to describe a circle eccentric to the earth, and how he determined the magnitude and direction of the eccentricity (Chapter I, § 1, *supra*).

Copernicus, as we shall see, distinguishes further causes of non-uniformity in the Sun's apparent motion, but he retains the eccentric to represent this fundamental inequality, subject to an interchange of the parts played by earth and sun. He reproduces the calculation of the elements of the eccentric given in the *Almagest* and then, having noted that several medieval observers disagreed with Ptolemy's estimates, he redetermines these quantities from observations of his own, made in 1515, the procedure being, briefly, as follows (III, 16):

Let ABC be the earth's orbit with center E (Fig. 15), G and L being the apses, at which the earth is respectively nearest to and farthest from the sun, F; and let A, B, C be the positions of

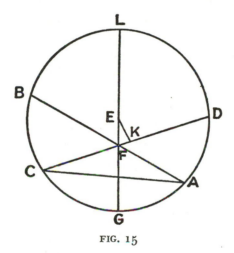

FIG. 15

the earth from which the Sun appears at the spring equinox, at the autumn equinox, and at the mid-point of the sign of Scorpio, respectively. Then the eccentricity (EF : EG) and the direction of the line LG have to be determined. The points A, F, B form a straight line; produce CF to meet the circle again in D, and draw EK perpendicular to CD.

Copernicus observed the time intervals required for the earth

to travel from B to C and from B to A. From tables of the mean
rate of motion of the sun in longitude (III, 14), he calculates the
arcs BC, BA, uniformly traversed by the earth during these in-
tervals. From the arc BC, thus known, the $\angle\,$BAC ($= \frac{1}{2}\,\angle\,$BEC)
is obtained; also $\angle\,$BFC is known ($= 45°$); hence $\angle\,$ACD
($= \angle\,$BFC $- \angle\,$BAC) is found, and the arc AD. Hence, and
from the known arcs BC and BA, the arc CAD is obtained.
From the Table of Chords, Copernicus finds the ratio of chords
(CD : CA), and also, in the triangle ACF, the angles of which
are all known, he finds (CF : CA), and hence (CF : CD),
(CF : FD), and, eventually, FK (which is half of the difference
of FD and CF) in terms of the diameter of the circle ABC. The
perpendicular EK is found in the same units, from $\angle\,$EDK,
which is half of ($180° - \angle\,$CED); hence the right-angled triangle FKE can be solved. The eccentricity (EF : EG) of the earth's
orbit is found to be about $\frac{1}{31}$, and the sun's apogee is found to
lie $6°\ 40'$ east of the summer solstitial point.

It is now possible to calculate, for any given day in the year,
the sun's *equation of center,* i.e., the difference between the
mean and the apparent places of the sun in the ecliptic, which
arises from the eccentricity of the earth's orbit (III, 17). For let
AEB (Fig. 16) be the earth's eccentric, center C, with the sun at

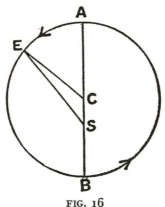

FIG. 16

S. To an observer on the earth, say at E, the apparent direction of the sun (ES) differs from its mean direction (EC, which appears to revolve uniformly about E) by the angle CES. This angle can be readily calculated from a knowledge of the earth's elongation (ACE) from the farther apse A and of the eccentricity (CS : CE). The equation of center ∠ CES can be shown to be a maximum when ES is perpendicular to AB (III, 15; cf. *Alm.* III, 3). This maximum, as calculated by Ptolemy, was 2° 23′; with Copernicus' elements, it amounts to 1° 51′ (modern maximum value of the equation of center: 1° 55′).

Having found reason to suppose that the direction of the apse line of the earth's eccentric and the amount of the eccentricity are alike subject to change with lapse of time, Copernicus proceeds to investigate the law governing these variations (III, 20). His stock of observational data was small, and his theory is carefully adjusted to harmonize with Ptolemy's assertion that the eccentricity was the same in his time as in the days of Hipparchus. Copernicus supposes that the apse line swings round non-uniformly from west to east, and he assumes somewhat arbitrarily that the variations in its rate of motion and in the eccentricity occur in a cycle of roughly the same period as that governing the variations in the obliquity (§ 2, *supra*). He proposes two alternative, equivalent methods of representing these variations; we shall here describe the simpler of the two.

The center, C, of the earth's circular orbit, KLM (Fig. 17), describes a small circle, EF, from east to west about a point, G, which is fixed relatively to the sun, S. As C revolves, the apse line LSCK of the earth's orbit evidently oscillates about its mean direction, BSGA (which itself turns uniformly about S from west to east), while the eccentricity, SC, oscillates between its maximum and minimum values, SE and SF.

Second and third solar inequalities arise from these variations in the eccentricity and in the motion of the apse line, respectively; and Copernicus, as we have already noted, regards these

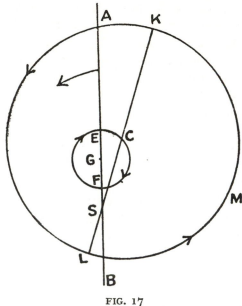

FIG. 17

as governed by the same anomaly as the obliquity. This assumption enables him readily to determine the mean apse line and the correct elements for computing the equation of center at any given epoch.

The anomaly goes through its cycle of 360° in 3434 years (§ 2, *supra*), which is therefore the period in which the center of the earth's eccentric revolves round its mean position. Taking the diameter of the earth's orbit as 10,000 parts, the eccentricity was 414 between the times of Ptolemy and Hipparchus (in which period the anomaly passed through its zero value; see § 2, *supra*) and 323 at the date (1515) of Copernicus' own observation (when the anomaly would be 165° 39′). From these data, the required elements are obtained as follows (III, 21):

Let EFD be the small circle described about G by the center of the earth's orbit (Fig. 18; cf. Fig. 17), and let S be the sun;

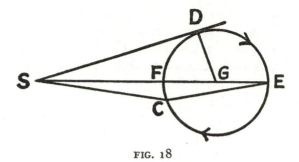

FIG. 18

then SGE is the direction of the mean apse line, and SE, SF measure, on a certain scale, the maximum and minimum eccentricities, respectively. From E, set off the arc EC subtending 165° 39' at G. Then Copernicus takes the direction of SC as that of the true apse line in 1515, and its length as measuring the eccentricity in that year. To calculate EF, SG, and $\angle$ ESC, we have

$$SE = 414 = \text{max. eccentricity; } SC = 323,$$

and $$\angle CES = 7° 10',$$

whence, by calculation, EC = 95, EF = 96,

whence SF = (414 − 96) = 318 = min. eccentricity,

$$SG = 366, \text{ and } \angle ESC = 2° 7';$$

this is the *equation of the apogee* (i.e., the difference between the longitudes of the mean and apparent apogees arising from the non-uniformity of motion of the apse line) in 1515.

Now draw SD to touch the circle at D, and join DG. Then

$$SG = 366, \text{ and } DG = 48,$$

whence $\angle$ DSG, the *maximum* equation of the apogee, is found to be 7° 28'.

The mean progressive motion of the apse line is deduced (III, 22) from a comparison of the approximate place of apogee in 64 B.C. (when the anomaly was supposed to have vanished and the mean and apparent apogees to have coincided) with its calculated mean place in A.D. 1515. The motion is found to be about 24″ per annum from west to east, relative to the fixed stars (modern value: 11″·25 *per annum*).

A correction has thus to be applied (III, 23) to the sun's mean rate of motion relative to the stars, so as to give its motion as measured from mean apogee; and a further correction is required, in calculating the equation of center, for the departure of the apparent from the mean apogee. These corrections are provided for in the tables by which Copernicus completes his solar theory (III, 24). These tables show

(*a*) the equation of the apogee corresponding to each 3° of the argument of anomaly;

(*b*) the sun's equation of center for each 3° of its elongation from the apparent apogee, when the eccentricity has its minimum value;

(*c*) the amounts to be added to the equations of center (*b*) when the eccentricity is a maximum;

(*d*) the proportions of the corrections (*c*) to be applied for each 3° of the anomaly.

Our study of the Copernican theory of the earth's motions, then, reveals the following innovations of importance to the progress of astronomy:

(*a*) The phenomena of the day and year are referred to a combined rotation and revolution of the earth.

(*b*) Precession is interpreted, for the first time, as due to a slow conical motion of the earth's polar axis about the axis of the ecliptic.

(*c*) The progressive motion of the apse line of the earth's orbit, detected by al-Battani (Chapter I, § 2, *supra*), is confirmed, and a reasonable estimate of its magnitude offered.

For the rest, Copernicus' detailed theories of the supposed fluctuations in the precession, the obliquity, the eccentricity, and the progression of the apse line are worthless for the reasons already indicated. Fluctuations in these quantities do indeed occur, and our methods of representing them bear, in some instances, a formal resemblance to the devices of Copernicus; but any comparison of our numerical coefficients with his would be meaningless.

5

The Copernican System:
Theory of the Moon's Motion

IN THE COPERNICAN SYSTEM, THE MOON RETAINED ITS FORMER RE-
lation to the earth, exchanging its status as one of the seven
planets for that of a satellite, the daily rising and setting of which
was now to be attributed to the earth's axial rotation. Hence the
substitution of the heliocentric for the geocentric point of view
did not fundamentally affect the problem of giving a geometrical representation of the moon's apparent motion; and the
fourth book of the *De revolutionibus,* which is devoted to the
lunar theory and related problems, is among the less original
portions of the work. Nevertheless, while adopting many of
Ptolemy's observational data and employing his characteristic
geometrical methods, Copernicus was able to make considerable
improvements upon the lunar theory set out in the fourth and
fifth books of the *Almagest.*

§ 1. THE LUNAR INEQUALITIES

Copernicus begins his study of the moon's motion by examin-
ing Ptolemy's lunar theory, the essential features of which we
have already outlined (IV, 1; cf. *Alm.,* V, 2, and Chapter I, § 1,
supra). He regards this theory as objectionable and inadequate
(IV, 2). For it was supposed that the center of the epicycle appeared to move uniformly with regard to the earth's center, and
this would mean that it must actually move non-uniformly about
the center of its own deferent, and this would conflict with the

fundamental axiom of celestial motion (I, 4). For the same reason Copernicus rejects the convention whereby the moon was supposed to move in its epicycle with a uniform angular velocity relative to a line joining the center of the epicycle to an arbitrary point within the deferent. He further declares that the classical lunar theory indicates fluctuations in the moon's distance and angular diameter which are out of all proportion to those actually observed.

With a view to removing these objections, Copernicus proposes to represent the moon's motion as follows (IV, 3):

He makes the moon describe a small epicycle, DF (Fig. 19), so

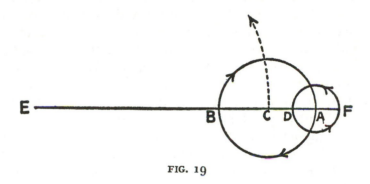

FIG. 19

that starting from F it moves from west to east; the center, A, of this circle describes, in the opposite sense, a larger epicycle, AB, of which the center, C, is carried from west to east on a deferent the center of which is that of the earth, E. When EC is directed toward the mean sun, the moon is to be at D on its epicycle; when EC has turned through 90° from this position, the moon is to be at F. It thus traverses the circle DF twice in a synodic month (mean interval between two successive new moons) relative to the line DC. Similarly, A describes the larger epicycle in an anomalistic month (interval between two successive passages of the moon through its perigee), while C describes its deferent (relatively to the mean sun) in one synodic month.

Copernicus provides tables (IV, 4; cf. *Alm.*, IV, 3) showing the accumulated angular motion of the moon from day to day and from year to year, relative (*a*) to the sun (synodic motion), (*b*) to the apses of the moon's orbit (anomalistic motion), and (*c*) to its nodes on the ecliptic (draconitic motion).

The next few chapters (IV, 5–12) are devoted to an investigation of the two principal lunar *inequalities,* or periodic fluctuations in the moon's rate of motion about the earth; they are represented by the two epicycles of Fig. 19, and they correspond respectively to the *equation of center* and the *evection* in modern lunar theory. Both inequalities were known to Ptolemy (*Alm.*, IV, 4); and no more were discovered until after the time of Copernicus. The first inequality is an oscillation of the moon about the mean position in its orbit, vanishing at apogee and at perigee, and recurring in the moon's anomalistic period. It arises in consequence of the eccentricity of the moon's orbit, and allowance was made for it by the Babylonian astronomers and by Hipparchus. The second inequality is here treated as a variation in the first one, disappearing about new and full moon, and becoming most pronounced at quadrature; it arises from fluctuations in the eccentricity of the moon's orbit, and it seems to have been recognized first by Ptolemy (*Alm.*, V, 2).

The problem before Copernicus was that of deducing from observations the ratios (CD : EC) and (CF : EC) in Fig. 19, so as to determine the elements of his lunar theory. He calculates the former ratio from a group of three lunar-eclipse observations, in imitation of Ptolemy's procedure for determining the first inequality (*Alm.*, IV, 5). He applies the method to two independent groups of three eclipses each. We shall pass over the discussion of the first triad of observations (those which Ptolemy had utilized) and shall sketch Copernicus' application of the method to three eclipse observations which he had made for himself (IV, 5). On each occasion he had recorded the sun's longitude and the time of the central phase of the eclipse to within a fraction of an hour, with the following results (we shall give merely the dates of the eclipses and the true positions of the sun in the zodiac):

DATE	TRUE SUN
(a) 1511, Oct. 7	22° 25′ of Libra
(b) 1522, Sept. 6	22° 12′ of Virgo
(c) 1523, Aug. 26	11° 21′ of Virgo

In the mean interval of 10 yrs., 337 d., 48 mins., between the eclipses (a) and (b), the moon evidently moved through 329° 47′, over and above complete circuits, since at each eclipse it was diametrically opposite to the sun. In the mean interval of 354 d., 3 hrs., 9 mins., between (b) and (c), the moon's motion was 349° 9′. For the former interval of time, the mean synodic motion of the moon relative to the sun, rejecting multiples of 360°, was given by the tables as 334° 47′, and the motion of the moon in anomaly (motion in the circle AB of Fig. 19, relative to the apogee) as 250° 36′. For the latter interval of time, the tabular synodic motion was 346° 10′ and the anomalistic motion, 306° 43′. Hence, in the former interval, a motion of 250° 36′ in anomaly had the effect of taking (334° 47′ − 329° 47′), or 5°, from the moon's mean motion; while in the latter interval a motion of 306° 43′ in anomaly added 2° 59′ to the moon's mean motion.

Now let E be the center of the earth (Fig. 20) and ABC the

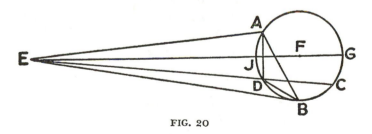

FIG. 20

circle upon which the moon lies when new or full, the center, F, of this circle being also the center of the moon's larger epicycle. Let A, B, and C represent the moon's positions on this circle at the first, second, and third eclipses respectively. Then, from the data given above,

$$\text{arc } ACB = 250° \ 36',$$
$$\text{arc } BAC = 306° \ 43', \text{ whence}$$
$$\text{arc } AGC = 197° \ 19', \text{ and}$$
$$\text{arc } \ \ CB = \ \ 53° \ 17' = \angle CFB;$$
$$\angle AEB = \ \ \ 5°, \text{ and } \angle BEC = 2° \ 59'.$$

Since the arc AGC exceeds 180°, and C is behind A in longitude, evidently the apogee must lie on AGC. Let G and J be the apogee and perigee respectively. Let EC intersect the circle ABC at D. Then $\angle CDB = \frac{1}{2} \angle CFB = 26° \ 38'$, and $\angle BED = 2° \ 59'$; therefore $\angle DBE = 23° \ 39'$, whence in the triangle BED, the Table of Chords gives $(BD : ED) = \frac{1042}{8024}$.

Likewise $\angle ADC = 98° \ 40'$, and $\angle AEC = 5° - 2° \ 59' = 2° \ 1'$, whence $EAD = 96° \ 39'$, and, in the triangle ADE, $(AD : ED) = \frac{702}{19865}$. But $(BD : ED) = \frac{1042}{8024}$ (proved above), whence $(AD : DB) = \frac{283}{1042}$. Also $\angle ADB = \frac{1}{2}$ (re-entrant) $\angle AFB = 125° \ 18'$, whence, solving the triangle ADB, $(AB : BD) = \frac{1227}{1042}$.

Expressed in terms of the radius of the circle ABC (FG = 10,000 parts),

$$AB = 16{,}323; \ BD = 13{,}853; \ ED = 106{,}751;$$

whence the arc DB subtends 87° 41′ at the center. Add the arc BC, and the arc DC will subtend 140° 58′ at the center. On the scale FG = 10,000, we have (from the Table of Chords) DC = 18,851 and EC = ED + DC = 125,602. Now

$$EJ \cdot EG = EF^2 - JF^2 = ED \cdot EC \ (\textit{Euclid}, \text{ III, } 36);$$

whence EF = 116,226, where FG = 10,000,
or   FG = 860, where EF = 10,000.

This is found to be in fair agreement with the value of FG (870) which Copernicus had deduced from Ptolemy's observations.

Copernicus proceeds to calculate the angles FEA, FEB, FEC, representing the differences between the apparent place of the moon and its mean place (F) at each of the three eclipses. Knowing what the apparent places were on these occasions, he is thus able to calculate the corresponding mean places. A comparison (IV, 6) of one of these mean places with one similarly deduced from Ptolemy's triad of observations enables him to verify the accuracy of the moon's rates of synodic and anomalistic motion assumed in the tables (IV, 4).

Copernicus has next to investigate the second lunar inequality. The problem presents itself in this way: It was found that when the moon was near quadrature (90° from the sun), the difference between its mean and its apparent positions might amount to as much as 7° 40′, although, if the moon were restricted to move on the circle ABC of Fig. 20, this difference should not exceed about 5° (IV, 8; cf. *Alm.*, V, 3). Hence the necessity for the second smaller epicycle in addition to the first.

About C, the mean position of the moon in its orbit (Fig. 21), draw the circle ADB upon which the moon lies at the times of its greatest departure from its mean position. Through E, the center of the earth, draw the line EBCA and the tangent ED meeting the circle ADB at D. Join CD. The ∠ CED being 7° 40′,

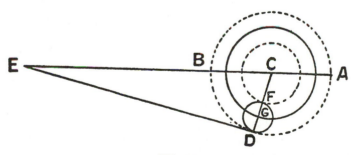

FIG. 21

and $\angle$ CDE being 90°, the ratio (CD : CE) is found to be $\frac{1334}{10000}$. But the moon's distance from the center C at full moon was found above to be about 860 parts (where CE = 10,000). Now draw a circle with center C and radius CF = 860; this is the circle upon which the moon lies at new and at full moon. Bisect FD at G, and draw a circle with center G and radius GF. This circle is the moon's second epicycle; its radius GF is ½ (1334 — 860), or 237 parts, the radius CG of the first, larger epicycle being 1097 parts. The angles subtended by these radii at the earth's center, E ($\angle$ CEG = 6° 18′, and $\angle$ GED = 1° 22′), correspond respectively to the amplitudes of the equation of center and of the evection (6° 17′, and 1° 16′) in our terminology.

This second inequality itself introduces a non-uniformity into the rate at which the moon revolves about the center C, of the larger epicycle, AB, relative to the apogee, A, i.e., the rate at which its anomaly increases (IV, 9). For let E be the center of the earth and D that of the smaller epicycle, and draw the tangents CL and CM to the latter (Fig. 22). It has been shown above

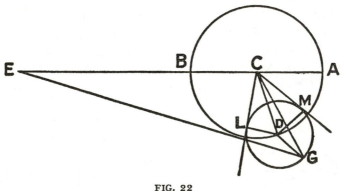

FIG. 22

that (DL : DC) = $\frac{237}{1097}$ = $\frac{2160}{10000}$, whence $\angle$ DCL = $\angle$ DCM = 12° 28′, and this is the greatest difference that can occur between the moon's mean and true anomalies. It will occur when

EC is inclined at an angle of $\frac{1}{2}$ (77° 32'), or 38° 46', to the line joining the center of the earth and sun.

Having arrived at a numerically determinate lunar theory, and having partly verified it (IV, 10) by reference to an observation of the moon ascribed to Hipparchus (*Alm.*, V, 5), Copernicus next proceeds to draw up a table by means of which the moon's apparent place can be deduced from its mean place at any given date (IV, 11, 12). The construction of the table depends upon the following principle: Let G be the moon (Fig. 22). Then in the triangle CDG the sides CD and DG, are constant and ∠ CDG is determined at any time by the moon's mean elongation from the sun. Hence CG and ∠ DCG may be found. Add ∠ DCG to the moon's mean elongation from perigee (∠ ECD), and the ∠ ECG is obtained; this, with EC and CG, enables the triangle ECG to be solved and ∠ CEG, the difference between the moon's mean and apparent places, to be obtained.

§ 2. THE MOON'S MOTION IN LATITUDE

The cycle of changes in the moon's celestial latitude recurs in the period required for the moon to perform a complete revolution relative to the nodes in which its orbit cuts the ecliptic. This is the *draconitic month,* to the evaluation of which Copernicus next addresses himself (IV, 13).

The ideal method of determining this period would, Copernicus holds, be to consider two partial lunar eclipses, occurring near the same node, showing shadows equal in extent and similar in situation, and falling on dates which, although widely separated in time, correspond to equal elongations of the moon from its apogee (cf. *Alm.*, IV, 2). The interval between two such eclipses would be that required for a whole number of circuits relative to the node, the particular whole number being given by a previous approximate knowledge of the required period.

Having failed to find such a convenient pair of eclipses, Copernicus resorts to another method and considers two eclipses occurring at nearly diametrically opposite points of the moon's orbit (relative to the line of nodes), the shadows encroaching from opposite points of the moon's limb and differing in extent by a known amount. The dates, which fell in 174 B.C. and A.D. 1509 respectively, corresponded to approximately equal distances of the moon from the earth. Copernicus finds the *apparent* motion of the moon, relative to a node, in the interval by allowing for the difference in the extents of the eclipses (a difference of $\frac{1}{12}$ of the moon's diameter in the extent of the shadow being taken to correspond to $1/2°$ difference of angular distance from the node); he converts this into *mean* motion and utilizes the result to verify his tables of the moon's motion in latitude (given in IV, 4).

§ 3. THE DISTANCES OF THE SUN AND MOON FROM THE EARTH

The earliest known attempt to determine the distances of the sun and moon from the earth by rigorous geometrical reasoning from the results of actual observations was that made by Aristarchus of Samos (*fl. c.* 280 B.C.). Aristarchus considered the triangle having the sun (S), the moon (M), and the earth (E) at its three angles. When the moon appeared half illuminated, he estimated the $\angle$ SEM to be 87°, the angle SME being then 90°; thus the three angles of the triangle, under these circumstances, were known, and Aristarchus was able to approximate to the ratio (ES : EM) which gave the relative distances of the sun and moon from the earth. (That he found this ratio to lie between 18 and 20, whereas actually it is about 380, was due not to errors of reasoning but to the fact that, at half-moon, $\angle$ SEM is about 89° 50′ and not 87°.) From further observational estimates of the angular diameters of the sun and moon and of the cross section the earth's shadow-cone at the moon's distance, Aristarchus was able to express the distances and sizes of the sun and moon in terms of the length of the earth's radius (although far from accurately). The earth's radius was itself determined in standard units of length by the ancients on more than one occasion (e.g., by Eratosthenes of Alexandria in the third century B.C.). They employed a method of measuring the size of the earth which has persisted, with increasing refinement, until modern times. The procedure was to find the difference of latitude between two stations lying on the same meridian and a known distance apart; this distance stood to the entire circumference of the earth in the same proportion as the observed difference of latitude stood to 360°.

Aristarchus had proceeded on the assumption that the size of the earth was negligible in comparison with the moon's distance from the earth. In actual fact, however, the distance of the moon is only about sixty times the earth's radius, and, in consequence, the direction in which a terrestrial observer sees the moon's center at some given instant varies appreciably according to his position on the earth. The *parallax,* or difference in apparent direction, may amount to as much as about 1°, which was an angle easily measurable by the observers of the ancient world; and it was by the measurement of such parallax that the moon's distance was eventually determined. The nearer the moon is to the observer's horizon, the more its apparent position among the surrounding stars differs from its position as seen, at the same instant, from a place where it is vertically overhead. But, given the moon's altitude above the horizon, the amount of the parallax depends simply upon the proportion of the earth's radius to the moon's distance. Hence Ptolemy was able to determine this proportion by comparing observations of the moon made when it was near the zenith with others made when it was near the horizon (*Alm.,* V, 13). Ptolemy gives the apogee distance of the moon as $64\frac{1}{6}$ times the earth's radius, and he calculates the corrections to be applied for parallax, so as to convert the observed position of the moon among the stars into what its position would appear to be if it were vertically overhead, for various positions of the moon in its orbit.

Copernicus rejects Ptolemy's estimate of the moon's distance and his system of corrections for parallax as resting upon a lunar theory which has been superseded by his own (IV, 16). He bases his own estimate upon two observations which he made, in 1522 and 1524, of the moon's meridian zenith distance (angular distance from the zenith when crossing the meridian). The differences between the observed and the calculated zenith distances on these two occasions represent the respective parallaxes, and they can be used to determine the corresponding distances of the moon by the following method (IV, 17):

Let ABA' be a section through the center, C, of the earth (Fig. 23). Let AD be the vertical of the observer A, and let E be the

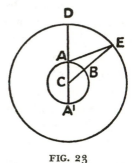

FIG. 23

moon's center. From the observed zenith distance, ∠DAE, of the moon, as seen from A, and the parallax, ∠AEC, by which it differs from the calculated zenith distance, the angles of the triangle AEC are known, and (CE : AC) can be calculated, giving the moon's distance in terms of the earth's radius at the time of the observation. To arrive at the greatest and least possible distances of the moon from the earth, let ABC (Fig. 24) be

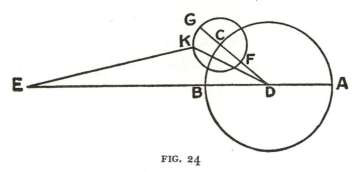

FIG. 24

the moon's larger epicycle with center D, FGK the smaller epi-cycle with the moon at K, and E the earth's center.

The moon's true anomaly, $\angle$ ADK (re-entrant), and hence $\angle$ EDK, are obtained, for the date of observation, from the tables, and also the difference $\angle$ DEK between the mean and apparent places of the moon in its orbit. Hence the angles of the triangle KDE are known, and Copernicus finds the ratio (DE : EK). But he has already found the moon's distance, EK, in terms of the earth's radius; and the elements of his lunar theory (IV, 5, 8) give DF and DG in terms of DE. Hence all these lengths can be expressed in terms of the earth's radius (and its sexagesimal fractions), the results being

$$\text{EK} = 56\tfrac{42}{60}, \qquad \text{DF} = 5\tfrac{11}{60},$$
$$\text{DE} = 60\tfrac{18}{60}, \qquad \text{DG} = 8\tfrac{2}{60};$$

whence the moon's greatest distance from the earth $= 68\frac{1}{3}$, and the moon's least distance from the earth $= 52\frac{4}{15}$, in the same units.

In his attempt to calculate the distance and size of the sun and the extent of the earth's shadow-cone (IV, 18–20), Coper-nicus follows the method described by Ptolemy (*Alm.*, V, 15) and attributed by him to Hipparchus. This method may be briefly indicated as follows:

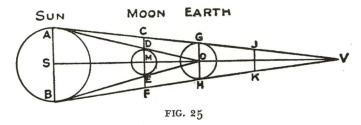

FIG. 25

In Fig. 25, let S, M, and O be the centers of the sun, the moon (both at apogee), and the earth respectively, when they are in the same straight line, and let AGV and BHV be common tangents to the sun and earth in the plane of section, intersecting in V, the vertex of the cone of shadow cast by the earth. The method presupposes that we know the following quantities: (a) the approximate distance OM of the moon from the earth in terms of the earth's radius OH; (b) the (equal) angular diameters, $\angle AOB$, $\angle DOE$, of the sun and moon, as observed from the earth; (c) the ratio in which the diameter JK of the earth's shadow at the moon's distance stands to the diameter DE of the moon (Ptolemy takes this ratio to be $\frac{13}{5}$; Copernicus makes it $\frac{403}{150}$). Knowing the moon's distance OM in terms of OH, and its angular diameter $\angle DOE$, we have its linear diameter DE. Knowing DE and (JK : DE), we have JK. Also, since JK and CF are equidistant from O, JK, GH, and CF are in arithmetical progression, so that CF = 2 GH — JK, and is known in terms of OH. Since EF = ½ (CF — DE), the ratio (EF : OH) is known. But

$$\frac{EF}{OH} = \frac{MS}{OS}, \text{ and } 1 - \frac{EF}{OH} = 1 - \frac{MS}{OS} = \frac{OM}{OS} = \frac{\text{moon's distance}}{\text{sun's distance}}$$

The sun's distance is thus known in terms of the moon's distance and hence in terms of the earth's radius. Ptolemy gave the ratio (sun's distance) : (earth's radius) as 1210. Copernicus, re-

I. Torun in the Time of Copernicus

(From L. Prowe, *Nicolaus Coppernicus*, 1883, 1884. Courtesy of the British Museum.)

II. Heilsberg

III. Heilsberg Castle

IV. FRAUENBURG

V. ALLENSTEIN CASTLE

(Plates II, III, IV, and V are from F. von Quast, *Denkmale der Baukunst in Preussen*, 1852–64. Courtesy of the British Museum.)

working Ptolemy's calculation with slightly different data, found for it a mean value of 1142; but his solar theory indicated that it should vary between 1105 and 1179. (The correct value would be about 23,000; the method is liable to give very inaccurate results, since the slightest error in estimating the acute and nearly equal vertical angles of the shadow-cones AOB and AVB greatly affects the resulting estimate of the distance of the sun, to which both these cones are tangential.)

In the next few chapters (IV, 21–23) Copernicus deduces the limiting values of the horizontal parallaxes and angular diameters of the sun and moon from the previously determined constants of their orbits; and he investigates the effect, upon the cross section of the earth's shadow at the moon's distance, of variations in the distances of the sun and moon from the earth. His next problem is that of tabulating the corrections for the parallax in altitude of the sun and moon (IV, 24, 25). Here again the work is modeled on the *Almagest* (*Alm.*, V, 17), but with such alterations as the new lunar theory requires. The remaining chapters of Book IV (26–32), which deal descriptively with such matters as eclipse problems, are not of such interest or originality as to call for discussion here.

The lunar theory of Copernicus represents no advance in fundamental principles over that of Ptolemy, for it recognizes no inequalities in the moon's motion beyond those already known. It had the merit, however, of representing the apparent motion of the moon without grossly exaggerating the observed variations in its angular diameter, as Ptolemy's theory had done. The reform which Copernicus effected in this respect may be judged from the following table showing the limits within which the angular diameter varies according to the respective theories of Ptolemy and Copernicus, with the limits given by modern observation, for purposes of comparison:

	Moon's Angular Diameter	
	Max.	Min.
Ptolemy	1° 0′ 26″	31′ 36″
Copernicus	37′ 34″	28′ 45″
Modern values	33′ 30″	29′ 26″

6

The Copernican System: Theory of the Planetary Motions

AS WE HAVE SEEN, COPERNICUS CLAIMS IN THE EARLY PAGES OF the *De revolutionibus* that his new theory can be harmonized with the facts of planetary motion and that it affords a simple explanation of certain well-known planetary phenomena. In the concluding two books of his work he seeks to justify this claim by constructing numerically determinate geometrical theories to represent the motions of the five planets, Mercury, Venus, Mars, Jupiter, and Saturn.

§ 1. SIDEREAL AND SYNODIC MOTIONS OF THE PLANETS

A distinction is drawn at the outset (V, 1) between the two components in the apparent motion of a planet in longitude. On the one hand, there is the orbital motion which the planet would be seen to possess by an observer stationed at the sun, whereby it completes a circuit of the heavens in its sidereal period. On the other hand, this motion is complicated for the terrestrial observer by his own motion on the moving earth, which gives rise to an apparent or parallactic motion (*motus commutationis*) of the planet, superimposing upon its orbital motion an inequality which recurs in the planet's synodic period. In consequence of this second component, a superior planet's apparent motion through the constellations is arrested and temporarily reversed at regular intervals. We have already referred to this phenomenon (pp. 20, 25) and have explained how an-

131

cient astronomers sought to account for it by postulating special spheres (Eudoxus) or epicycles (Ptolemy) as essential constituents of their planetary hypotheses. This had to be done separately for each planet; and it was found impossible to explain, on physical grounds, why the planetary motions should be complicated in this way. Copernicus, however, was able (V, 3) to account for this phenomenon in all the planets, as a class, by showing it to be a direct consequence of the annual revolution of the earth round the Sun—a single antecedent factor. This simple explanation of the peculiar behavior of the planets marked an important advance over the *ad hoc* hypotheses of Eudoxus and Ptolemy. It constituted the chief argument for the scientific truth of the Copernican theory and its best claim to acceptance when it was first put forward.

In order to understand, in a general way, how the retrogressions arise in consequence of the annual motion of the earth, suppose the respective orbits of the earth, E, and of a superior planet, P, to be coplanar circles having a common center at the sun, S (Fig. 26). Draw the tangents PE_1 and PE_2 from the planet to the earth's orbit. If the planet remained at rest at P and the earth uniformly described its orbit, then, while the earth tra-

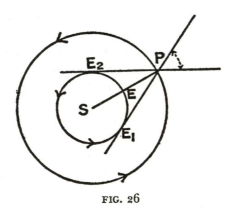

FIG. 26

versed the arc E_1EE_2, the planet would appear to a terrestrial observer to be moving against the background of stars in the retrograde direction (from east to west) through the $\angle E_1PE_2$; but while the earth described the remainder of its orbit, from E_2 to E_1, the planet would appear to move progressively (from west to east) through the same $\angle E_2PE_1$. That is, the planet would appear to oscillate about the direction SP with an amplitude $\angle SPE_1$, either way. But actually the planet P is not at rest, but travels round its orbit in the same direction as the earth (although with less angular velocity), and it carries the tangents PE_1, PE_2 with it. Thus the oscillations are superimposed upon a general eastward motion of the planet, and they give rise to the characteristic fluctuations in its apparent motion. The corresponding phenomena in the apparent motion of an inferior planet may be explained on analogous lines, the positions of the earth and planet in Fig 26 being interchanged. The period of recurrence of this inequality in a planet's motion is that in which the earth gains a complete revolution upon the planet, or vice versa; that is to say, it is the synodic period. Copernicus gives tables (based on Ptolemy's tables) showing how the synodic motion of each planet (its angular motion reckoned from the moving line SE) accumulates from day to day and from year to year (V, 1; cf. *Alm.*, IX, 4).

A numerically precise theory of this phenomenon subsequently given by Copernicus (V, 35, 36) is adapted from the investigation of the stationary points on an epicycle given by Ptolemy (*Alm.*, XII, 1) and attributed by him to Apollonius. The procedure is equivalent to determining the condition that the speeds of the earth and of a planet in the direction perpendicular to the line joining them should be instantaneously equal; the planet will then have no angular velocity about the earth and will appear stationary to a terrestrial observer.

§ 2. MOTIONS OF THE PLANETS IN LONGITUDE

Before embarking on the development of his own planetary theory, Copernicus gives a brief account (V, 2) of the hypothesis by which Ptolemy sought to represent the motions of Saturn, Jupiter, Mars, and Venus (*Alm.*, IX, 5, 6; see Chapter I, § 1, *supra*). The essential feature of Ptolemy's treatment was the epicycle carried round on a deferent eccentric to the earth. Copernicus stresses the fact that the center of this epicycle was supposed to move uniformly, not about the center of the deferent nor about the earth's center but about a third, arbitrarily chosen point, while the motion of the planet on its epicycle was measured from a radius passing through the same arbitrary point. He claims that his theory renders unnecessary such a travesty of the established principles of physics. Having shown (§ 1, *supra*) how the earth's annual revolution accounted, at one stroke, for the most conspicuous inequality in the apparent motion of each planet, he was free to concentrate on the residual effects arising from the fact that the planets do not uniformly describe circles concentric and coplanar with that traversed by the earth. The planets did not all yield to the same mode of treatment, three different types of hypotheses being required in order to represent the motions in longitude of (*a*) the superior planets, (*b*) Venus, and (*c*) Mercury, respectively. We shall now examine these several hypotheses and shall indicate how the constants which they involved were numerically determined from suitably chosen observations.

(a) The Superior Planets

The orbit of a superior planet was constituted as follows (V, 4):

The planet's deferent is the circle AGB (Fig. 27) with center D, and the earth's orbit is the circle NO with center C (the eccentricity being exaggerated in the figure for the sake of

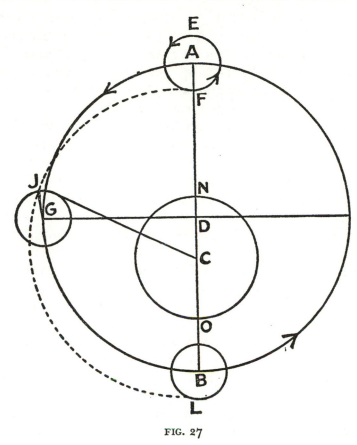

FIG. 27

clearness). The planet describes the epicycle EF, the radius of which is one-third of CD, uniformly from west to east, in the planet's sidereal period (relative to the moving radius of the deferent), while the center of the epicycle describes the deferent AGB in the same sense and in the same period. When the epicycle is at the farther apse, or apogee, A, the planet is at F; it is at J when the epicycle is at G (90° from A), and at L when the epicycle is at the nearer apse, or perigee, B. (It is worth noting that the curve FJL, described by the planet in a half-period, is not an exact semicircle.) By postulating this law of motion for a superior planet, Copernicus (without himself explicitly making the physically objectionable assumption) was yet able to obtain the advantage which Ptolemy had derived from assigning to the planet a uniform angular velocity about a point having a displacement from the center of its deferent equal but opposite to that of the earth (see Fig. 3, *supra*). It can, in fact, be shown that in the Copernican scheme the planet revolves uniformly about a point on DA distant from D by an amount equal to the radius of the epicycle. If G be the position of the center of the epicycle at any given time, and J the corresponding position of the planet, then the mean motion of the planet in longitude is measured by the increase in the angle ADG, and its apparent motion, as observed from the center of the earth's orbit, by the increase in the angle ACJ.

When the earth is exactly interposed between the planet and the center, C, of the earth's orbit, the planet's apparent place, as viewed by a terrestrial observer, is unaffected by parallax due to the displacement of the observer from C. Hence, whenever the planet appears in opposition to the center of the earth's orbit, the angle ACJ can be obtained free from error. The apse lines of the deferents of the superior planets were determined by Ptolemy from opposition observations (*Alm.*, X, 6); and Copernicus also determines the constants of his planetary orbits from groups of three opposition observations. He goes through the process twice for each of the three superior planets, first using three of Ptolemy's observations and then three of his own and proving that the two sets of results so obtained show a measure of agreement. We can sufficiently illustrate the procedure by considering a single determination of the orbit of a single superior planet, and we shall select Copernicus' discussion of his own observations of Saturn (V, 6). The dates of the three

oppositions and the longitudes of the planet on these occasions
were as follows:

DATE (A.D.)	LONG. OF THE PLANET
(a) 1514, May 5	205° 24′
(b) 1520, July 13	273° 25′
(c) 1527, Oct. 10 *	0° 7′

The changes in the planet's *apparent* longitude between (a) and
(b) and between (b) and (c), are evidently 68° 1′ and 86° 42′ respectively; and the *mean* motions in longitude corresponding to
the intervals of time between these two pairs of observations are
found to be 75° 39′ and 88° 29′.

The first step is to determine the eccentricity of the deferent
and the orientation of the apse line (AB in Fig. 27). To obtain
a first approximation, Copernicus, following Ptolemy (*Alm.*,
X, 7), assumes that the planet moves uniformly in a simple
eccentric, ABC, the center of which is F, that of the earth's orbit
being D (Fig. 28). His procedure is then as follows:

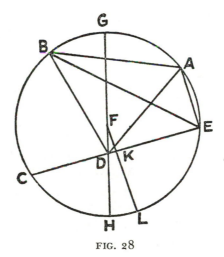

FIG. 28

* The date of the third opposition appears to be a mistake for 1527, Nov. 10.

Let A, B, C be the positions of the planet on this eccentric at the times of the first, second, and third oppositions respectively. Join AD, BD, CD, and AB; produce CD to meet the circle in E, and join AE, BE. From the foregoing data we have $\angle BDC = 86° 42'$ and $\angle BFC = 88° 29'$; hence $\angle BDE = 93° 18'$ and $\angle BED = \frac{1}{2} \cdot \angle BFC = 44° 14'$. Hence in the triangle BDE the remaining angle DBE is known, and from the Table of Chords the ratio of the sides BE, DE can be determined. Similarly, in the triangle ADE the ratio (AE : DE) and $\angle AED$ can be evaluated, and hence also the ratio (AE : BE). Knowing this ratio and $\angle AEB$ $(= \frac{1}{2} \cdot \angle AFB = \frac{1}{2} \cdot 75° 39')$, Copernicus solves the triangle AEB and obtains the ratio (AB : BE). But since $\angle AFB$ is known $(75° 39')$, AB can be expressed in terms of the radius of the circle ABC, and hence also BE and DE can be obtained in the same units. The value of BE found in this way shows that $\angle BFE = 103° 7'$, whence the re-entrant $\angle CFE$ amounts to $(103° 7' + 88° 29')$ or $191° 36'$; this enables the value of CE to be calculated, and hence the value of CD $(= CE - DE)$, in terms of the radius of the circle ABC. The center, F, of the circle must lie in the larger segment EABC. Let GFDH be the required apse line of the eccentric, and draw FK perpendicular to CE. We have

$$\text{(rectangle } CD \cdot DE) = \text{(rectangle } GD \cdot DH)$$
$$= \text{(square on } GF) - \text{(square on } FD).$$

But CD and DE are known in terms of the radius GF of the eccentric, and hence FD is obtained (FD = 1200 parts, where GF = 10,000; this is not far from the value of FD, viz., 1016 parts, which Copernicus had already deduced from Ptolemy's data). In the right-angled triangle DFK, DK is known $(\frac{1}{2} \cdot CE - CD)$, and so is FD, whence $\angle DFK$, or $\angle HFL$, can be calculated, and hence $\angle CFH$ $(= \frac{1}{2} \cdot \angle CFE - \angle HFL)$ and its supplement $\angle CFG$ are obtained. But $\angle CFB$ is known $(88° 29'$, as given above); $\angle BFG$ is found by subtraction, and hence $\angle GFA$

($= 35° 36'$) is obtained; this fixes the position of the apogee G
in relation to the positions of the planet at the three oppositions.
(Copernicus noticed that the longitudes of the apses of the three
superior planets seemed to have increased appreciably since the
time of Ptolemy, the apse line of Saturn, for example, having
swung round through about 14° during the intervening fourteen centuries. Such progressive motions of the apse lines actually occur, although Copernicus overestimated their amounts.)

The planet's orbit is now constituted as in Fig. 29 (cf. Fig. 27),

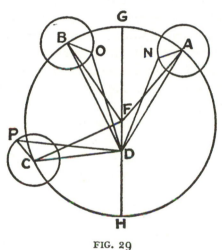

FIG. 29

where, as before, G and H respectively represent the apogee and
perigee; F and D are the centers of the orbits of the planet and
the earth respectively; and A, B, C, the planet's mean places at
the times of the three oppositions discussed above, are now to
represent the center of the planet's epicycle at those times. Of
the eccentricity (1200 parts) just deduced, three-quarters (900
parts) are assigned to the deferent to represent the separation
of the centers of the orbits of the planet and the earth; the remaining quarter (300 parts) appears as the radius of the epi-

cycle. The points N, O, P on the epicycle, where the planet is situated at the times of the respective oppositions, are defined, as in Fig. 27, by the relations

$$\angle FAN = \angle GFA;$$
$$\angle FBO = \angle GFB;$$
$$\angle FCP = \angle GFC.$$

The angles of *mean* motion, $\angle AFB$, $\angle BFC$, are known from the tables; and it is required to calculate the angles NDO and ODP and to show that they agree with the *observed* differences in the longitudes of the planet at the respective oppositions. Copernicus found that it would be necessary to adjust the calculated eccentricity and apse line somewhat in order to obtain such agreement, probably in consequence of the error in the date of the third opposition. He adopts the following corrected values:

$$\angle GFA = 38° \ 50';$$
$$FD = \quad 854 \text{ parts;}$$
$$\text{radius of epicycle} = \quad 285 \text{ parts,}$$
$$\text{where} \quad GF = 10,000 \text{ parts.}$$

Agreement between the observations and theory is then established as follows:

In the triangle AFD, AF = 10,000 parts; FD = 854, and $\angle AFD = 141° \ 10'$. Solving the triangle, we have AD = 10,679; $\angle FAD = 2° \ 52'$, and $\angle FDA = 35° \ 58'$.

In the triangle ADN, $\angle FAN = \angle GFA = 38° \ 50'$, whence $\angle DAN = \angle FAN + \angle FAD = 41° \ 42'$. Also AD and AN are known, whence $\angle ADN$ is found to be $1° \ 3'$. But $\angle ADF = 35° \ 58'$, therefore $\angle NDF = \angle ADF - \angle ADN = 34° \ 55'$. Similarly, from the second observation, $\angle ODF = 33° \ 5'$, whence $\angle ODN = 68°$, as against the *observed* difference of $68° \ 1'$ between the apparent longitudes of the planet at the first and second oppositions.

In like manner, $\angle ODP$ is found to be $86° \ 42'$, which agrees with the observed difference between the apparent longitudes at the second and third observations.

The longitude of the apogee G is found by subtracting from the observed longitude of the planet at the third opposition $(0° \ 7')$ the angle GDP $(= \angle ODF + \angle ODP = 33° \ 5' + 86° \ 42' = 119° \ 47')$; this gives the longitude of G as $240° \ 20'$.

It is thus shown that, so far as this triad of observations is concerned, the hypothesis enables arcs of mean motion to be correctly transformed into arcs of apparent motion. The hypothesis had been equally successful with Ptolemy's triad (V, 5); it is next shown to be applicable to the remaining superior planets (V, 10–19), and Copernicus feels confident in basing his Planetary Tables upon it.

We have seen that Copernicus determined the orbits of the superior planets according to his hypothesis, from observations of these bodies made when they were in mean opposition (i.e., when they appeared at an elongation of 180° from the mean sun at the center of the earth's orbit). In these circumstances they would appear, to a terrestrial observer, to lie in the same direction as if viewed from that center. But a superior planet may be observed at *any* elongation from the mean sun, and its direction will, in general, be affected by a parallactic displacement depending upon the relative positions of the earth and the planet in their orbits: this must be considered in calculating the planet's apparent (geocentric) longitude from its mean longitude, referred to the center of its own orbit. Moreover, a knowledge of the amount of this displacement, under given circumstances, enables the dimensions of the planet's orbit to be evaluated in terms of the radius of the earth's orbit. A single observation of the planet when it is not in opposition suffices for this purpose, assuming the motion to be in accordance with the hypothesis just described; and we shall again illustrate Copernicus' procedure with reference to the planet Saturn (V, 9).

At a certain hour on February 25, 1514, Saturn was observed to be in longitude 209°. The tables give the sun's mean longitude at that time (315° 41′) and Saturn's mean elongation from the mean sun* (116° 31′) and, hence, the planet's mean longitude as the difference of these (199° 10′), the apogee of its deferent being at about 240° 20′, as already stated.

In Fig. 30, let F be the center and G the farther apse of the

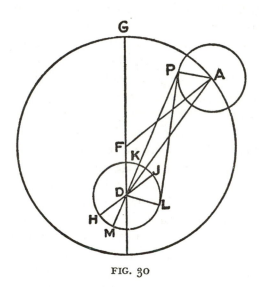

FIG. 30

planet's deferent; let D be the center of the earth's orbit. Let P be the planet and A the center of its epicycle, where AP = ⅓ · DF and ∠FAP = ∠AFG = (240° 20′ − 199° 10′) = 41° 10′, in accordance with the hypothesis already explained. Let PD produced cut the earth's orbit at K and M; and draw the diameter HJ parallel to FA. Set off the arc HL, making ∠HDL = 116° 31′; then L will be the position of the earth in its orbit. In the triangle AFD, the ratio (FD : AF) is known from the theory

* That is to say, the angle between the line drawn from the earth to the center of its orbit and the line drawn from the center of Saturn's epicycle to the center of its deferent.

of Saturn's motion, already reviewed, and $\angle AFD = 180° -$ $\angle AFG = 138° 50'$. Hence the triangle can be solved for AD and for $\angle DAF$ (which is $3° 1'$). Next, in the triangle PAD, AP, AD, and $\angle DAP (= \angle DAF + \angle FAP)$ are known, and the triangle can be solved for DP and for $\angle ADP$ (which is $1° 5'$). The planet's mean place is defined by the direction FA and its apparent place, referred to D, by the direction DP, and the difference between these is given by ($\angle ADP + \angle DAF$), or $4° 6'$. Hence, if the earth were at K or M, Saturn's apparent longitude would be ($199° 10' + 4° 6'$), or $203° 16'$. But, the earth being at L, the planet's longitude was observed, as has already been stated, to be $209°$. The difference ($\angle DPL$) of $5° 44'$ must be due to parallax. Now $\angle PDL = 180° - \angle LDM$, and $\angle LDM = (116° 31' - 4° 6') = 112° 25'$ (since $\angle HDM$ measures the inclination of FA to DP), whence $\angle PDL = 67° 35'$. Also $\angle DPL = 5° 44'$. Hence, in the triangle DPL, the ratio (DP : DL) can be calculated. But (AD : DP) is known from the triangle PAD, and (AF : AD) from the triangle FAD. Hence, eventually, the ratio (AF : DL) is obtained, i.e., the ratio of the radius of the planet's deferent to that of the earth's orbit. Thus, if DL be taken as unity, AF amounts to 9.174. The eccentricity DF and the radius AP of the epicycle can also be evaluated in the same units.

By analogous investigations of the radii of the other planetary deferents, Copernicus arrives at results which may be tabulated as follows: the modern "mean distances" of the several planets are given for general comparison with Copernicus' figures:

PLANET	RAD. OF DEFERENT	MOD. MEAN DISTANCE
Mercury	0.376	0.387
Venus	0.719	0.723
Earth	*1.000*	*1.000*
Mars	1.520	1.524
Jupiter	5.219	5.203
Saturn	9.174	9.539

We have here the earliest instance of an astronomer's determining the relative sizes of the orbits of the several planets from actual observations.

The various estimates made by astronomers before Copernicus of the relative distances of the planets from the center of the universe were all obtained by assuming these distances to be connected by some arbitrary relation. Thus the Pythagoreans supposed that the radii of the successive planetary orbits were proportional to the segments of a lyre-string the vibrations of which would give notes forming some musical concord. Plato, in the *Timaeus,* seems to have taken the distances of the seven planets from the center as respectively proportional to the numbers 1, 2, 3, 4, 8, 9, 27—a series formed by combining the first four terms of two geometrical progressions (1, 2, 4, 8 and 1, 3, 9, 27), and the last term of which is the sum of all the preceding ones. When it had been recognized that the distance of each planet from the earth varied within definite limits, certain late Greek and Arab astronomers made the assumption that the maximum distance of any given planet from the earth, at the center of the universe, was equal to the minimum distance of the planet immediately exterior to it, so that no useless empty spaces should exist in the universe. Now the greatest distance of the moon from the earth was known to the Alexandrians with fair accuracy, as we have seen (Chapter V, § 3, *supra*): it was assumed equal to the least distance of Mercury, the next planet beyond the moon in the Ptolemaic system. Moreover, knowing a planet's minimum distance from the earth, it was possible to calculate its maximum distance, given the ratio of the radii of the epicycle and deferent, and the eccentricity of the latter, which were both deducible from observations. Hence, the minimum distance of Mercury being assumed equal to the known maximum distance of the moon, the maximum distance of Mercury could be calculated; but this was the minimum distance of Venus. The maximum distance of Venus, similarly derived from the constants of its orbit, was equal to the minimum distance of the sun; and so on, up to the maximum distance of Saturn, which was generally assumed equal to the radius of the sphere of fixed stars. It happened that the distance of the sun obtained in this manner agreed fairly closely with that given in the *Almagest,* which was obtained by a totally different method. The procedure just outlined led to gross underestimates of the distance (and therefore of the magnitude) of the sun, and it gave an entirely false idea of the proximity of the stars; these results, combined with the absence of observable stellar parallax, would further tend to discourage heliocentric hypotheses during the Middle Ages.

We turn next to those chapters of the *De revolutionibus* which treat of the motions in longitude of the inferior planets.

(b) Venus

In his attempt to represent the motion of Venus, Copernicus assumes that, to a first approximation, the planet describes a circle about a center, D, which is displaced from the center, C, of the earth's orbit, AB (Fig. 31). This eccentric has an apse line ACDB, the orientation of which may be determined from ob-

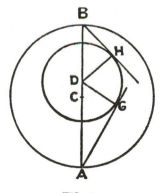

FIG. 31

servations of maximum elongations of the planet from the mean
sun, C, on the principle that, if two successive maximum elongations are observed to be equal, an apse must lie midway between the places of the mean sun at the respective observations.
A comparison of the amounts of different pairs of such equal
elongations serves to determine which is the farther and which
the nearer apse (V, 20; cf. *Alm.*, IX, 7, and X, 1). Following
Ptolemy's data and calculations (*Alm.*, X, 1), Copernicus locates
the farther apse in longitude 48° 20′ (that is, C appears from A
to be in longitude 48° 20′) and the nearer apse in longitude
228° 20′.

The radius and eccentricity of the planet's eccentric are
found (V, 21) by assuming, again on the authority of Ptolemy,
that the maximum elongation of the planet from C is 44° 48′
when it is observed from the earth at the farther apse and is
47° 20′ when observed from the nearer apse. Draw the tangents
AG and BH, and join DG and DH. In the right-angled triangle
ADG, $\angle$ DAG = 44° 48′, whence (DG : DA) is found to be
$\frac{7046}{10000}$. Similarly, in the right-angled triangle BDH, $\angle$ DBH
= 47° 20′, whence (DH : DB) = $\frac{7346}{10000}$. Taking DG = DH
= 7046, we have DB = 9582. Hence, AB = AD + DB =
19,582, whence AC = 9791, and CD = 209. Hence, on the
scale AC = 10,000, we have DG = DH = 7193 and CD = 213.
Thus, taking the radius of the earth's orbit as unity, the radius
of Venus' eccentric is 0.719.

Copernicus found, however, that this simple theory broke
down when the earth was not at an apse (V, 22). He takes the
following pair of observations (*Alm.*, X, 3) into consideration:

DATE	MEAN SUN	ELONGATION OF VENUS
(a) 134 A.D., Feb. 18	318° 50′	43° 35′ (morning star)
(b) 140 A.D., Feb. 18	318° 50′	48° 20′ (evening star)

At each observation the mean sun, C, appeared in a direction at right angles to the apse line ACB of Venus, the earth being at E (Fig. 32), where $\angle ACE = 90°$. The calculation is then as follows:

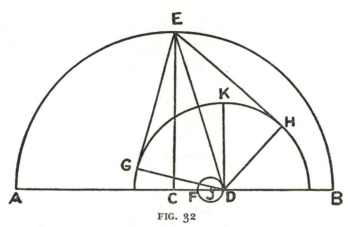

FIG. 32

Through the center, D, of the eccentric of Venus draw DK perpendicular to AB, and construct the tangents EG, EH. Then

$\angle GEC = 43° \, 35'$;

$\angle HEC = 48° \, 20'$;

∴ $\angle GEH = 91° \, 55'$, and $\angle DEH = \angle DEG = 45° \, 58'$;

∴ $\angle CED = 2° \, 23'$. But $\angle DCE = 90°$; hence the triangle CED can be solved, and (CD : CE) is found to be $\frac{416}{10000}$. But in the former theory, (CD : AC) was approximately $\frac{208}{10000}$. Now bisect CD at F, so that CF = FD = 208. Bisect FD at J. Copernicus regards the center of the eccentric of Venus as describing from west to east the circle FD about J, so that when the earth is at A or B the center of the planet's eccentric is at F, but when the earth is midway between the apses, as at E, the center is at D.

This hypothesis led to results in agreement with the ancient data; but contemporary observations seemed to indicate that, whereas FD had remained at 208, CD had diminished from 416 to 350.

(c) Mercury

The problem of representing the motion of the planet Mercury had taxed the ingenuity of Ptolemy, and the most complicated of all the schemes of planetary motion in the *De revolutionibus* is the one relating to this member of the solar system (V, 25). The orbit of Mercury, as it was conceived by Copernicus, may be explained by reference to Fig. 33.

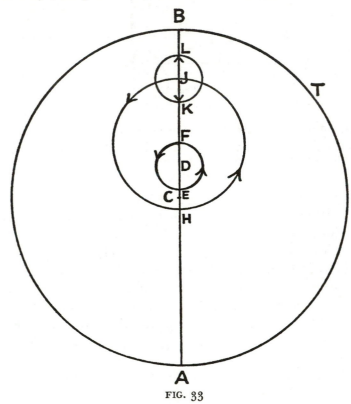

FIG. 33

The circle ATB is the earth's orbit, with center C. About D, a point on the diameter AB, a circle EF is drawn. With center F (the point on EF farthest from C) a circle HJ is drawn, which is to be the eccentric of Mercury; and with center J is drawn the epicycle LK.

The center F of the eccentric HJ is conceived to describe the circle EF from west to east at twice the earth's rate of revolution round the sun, and J swings round about the center F from west to east in the planet's sidereal period of 88 days, relative to the stars, or in its synodic period of 116 days, relative to the earth. The planet meanwhile oscillates between L and K on the diameter LK of the epicycle, with a motion compounded of circular motions, after the manner explained in connection with the theory of precession (Chapter IV, § 2, *supra*). It performs two complete oscillations in a year. When the earth is at one of the apses, A or B, the center of the planet's eccentric is at F and the planet is at K. When the earth is 90° from A and B, the center of the eccentric is at E and the planet is at L.

By the same method as that employed to determine the apse line of Venus' eccentric, Copernicus fixes the farther apse of Mercury in longitude 183° 20′ and the nearer apse in 3° 20′ (V, 26; cf. *Alm.*, IX, 7).

The numerical elements of Mercury's orbit are next determined (V, 27) from measurements of the maximum elongations of Mercury from the mean sun, with the earth (*a*) at an apse and (*b*) 90° from an apse:

Long. of Mean Sun	Max. Elongation of Mercury
(*a*) 182° 38′	19° 3′ (morning star)
4° 28′	23° 15′ (evening star)
(*b*) 93° 30′	26° 15′ (evening star)
93° 39′	20° 15′ (morning star)
(cf. *Alm.*, IX, 8, 9)	

The positions of the earth at the times of the observations (*a*) were nearly 180° apart and may be taken as coinciding sensibly with the apses. In Fig. 34, let C and D be the centers of

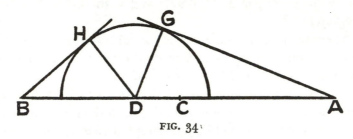

FIG. 34

the orbits of the earth and of Mercury respectively, and let A and B be the positions of the earth at the apses at the times of the respective observations (*a*). Draw tangents AG and BH to the planet's eccentric, and join DG and DH. From the observations (*a*),

$$\angle DAG = 19° \; 3', \text{ and } \angle DBH = 23° \; 15'.$$

Hence the ratios of the sides of the right-angled triangles AGD and BHD can be calculated, and, taking AC = 10,000, we have

$$DG = DH = 3,573, \text{ and } CD = 948;$$

this gives the radius and the eccentricity of Mercury's eccentric in terms of the radius AC of the earth's orbit.

Considering next the observations (*b*), we are to assume that the radius CE joining the earth, E, to the center of its orbit is perpendicular to the planet's apse line AB (Fig. 35) and that the center of Mercury's eccentric is now at some point F differing from the position D which it occupied (Fig. 34) when the earth was at an apse. Draw the tangents EG and EH to the planet's eccentric. Then Copernicus' problem is to find the position of F and to determine DF and FG. From the observations (*b*),

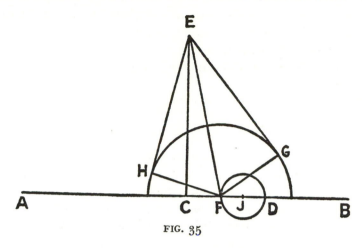

FIG. 35

$$\angle CEG = 26°\ 15',\ \text{and}\ \angle CEH = 20°\ 15'.$$
$$\therefore\ \angle GEH = 46°\ 30'.$$
$$\therefore\ \angle FEG = \angle FEH = 23°\ 15'.$$
$$\therefore\ \angle CEF = \angle CEG - \angle FEG = 3°.$$
Also, $\angle ECF = 90°$.

Hence the sides of the triangle ECF are found (in terms of CE = CA = 10,000) to be: CF = 524 and FE = 10,014. But CD was found above to be 948 on this scale. Hence FD = (948 — 524) = 424, which is the diameter of the small circle, FD, described by the center of Mercury's orbit. Hence, if J is the center of that circle, CJ = 736.

Also, in the right-angled triangle HEF, where $\angle HEF = 23°\ 15'$, FH is found to be 3953 where CE = 10,000. But when the earth was at A or B, it was shown (above) that the planet was at a distance 3573 from the center of its eccentric. Hence the amplitude of the planet's oscillation about its mean distance from the center must be (3953 — 3573) or 380.

The system of Mercury, as Copernicus conceived it, is thus numerically determined. He was able to prove its conformity with the results of observations of the planet made (in the years 1491 and 1504) by the Nuremberg astronomer Walther * (V, 30). The northerly latitude and humid atmosphere of Frauenburg prevented him from observing the elusive planet satisfactorily for himself.

§ 3. PLANETARY TABLES

Following on the development of his theory of the planetary motions, Copernicus furnishes a set of tables, with rules for their use (V, 33, 34); these make possible the calculation of the apparent longitude of a planet, at any given date, when its mean longitude is known. The quantities to be obtained from the tables are (a) the planet's *equation of center* (the difference between the planet's mean longitude and its apparent longitude as viewed from the center of the earth's orbit); and (b) the proper correction to be applied for parallax due to the displacement of the terrestrial observer from the center of the earth's orbit.

Copernicus' Planetary Tables, like all others of their kind until the seventeenth century, suffered from the inadequacy of the theory upon which they were based—a theory which was restricted to conform to illusory physical laws, and the constants of which were determined from a modicum of doubtful observations. A. F. Möbius sought to determine to what degree of approximation the geometrical scheme of planetary motion postulated by Copernicus could be adjusted to fit the actual motion of a planet. He showed that, if the planet's true anomaly be expressed as a series of terms involving the sines of successive multiples of the mean anomaly, then the construction of Copernicus was equivalent to neglecting all the terms containing

---

* Copernicus' ascription of two of these observations to Schöner was a mistake, according to Birkenmajer.

higher powers of the eccentricity than the first. He calculated that, in the case of Mars, for example, the resulting discrepancy in the heliocentric longitude of the planet might amount to 37′. This was on the assumption that the motion of the planet was referred to the true sun; referring it, as Copernicus did, to the center of the earth's orbit would increase the error to about 2°, or roughly twelve times the limit within which Copernicus sought to make his theory agree with observation, according to his disciple Rheticus, to whom he once said: "If only I can be correct to ten minutes of arc, I shall be no less elated than Pythagoras is said to have been when he discovered the law of the right-angled triangle" (Rheticus: *Ephemerides novae*, p. 6). And when the heliocentric longitude of the planet was converted into geocentric longitude, further errors arose from analogous inaccuracies in the representation of the earth's motion, so that the observed place of Mars might differ by as much as 3° from its tabular place. (For a synopsis of Möbius' calculations, see Apelt's *Die Reformation der Sternkunde*, pp. 261 ff.)

Some improvement in the accuracy and convenience of the Copernican tables was effected, a few years after their first appearance, by Erasmus Reinhold. The immediate gain to practical astronomers which resulted from the substitution of the heliocentric theory for the geocentric as the basis of planetary tables was, however, inconsiderable, and it might have been achieved merely by a refinement of the traditional tables. The significance of the Copernican theory lay not in the merits of the scheme of planetary motions initially associated with it, and soon to become obsolete, but in the fact that the adoption of the heliocentric standpoint was the indispensable precondition for the vast subsequent advances in astronomical theory which have made possible the precision of the lunar and planetary tables of today.

§ 4. MOTIONS OF THE PLANETS IN LATITUDE

Copernicus devotes the last and shortest book of the *De revolutionibus* to the problem of representing geometrically the observed departures of the planets from the plane of the ecliptic. The superior planets and the inferior planets form two groups, each requiring separate treatment; and we shall consider the two cases separately.

(a) Case of a Superior Planet

Copernicus represents the motions in latitude of the superior planets as follows (VI, 1): Each of these planets moves in a plane inclined to the ecliptic; it thus shows a maximum northward displacement from the ecliptic at one point of its orbit and a maximum southward displacement at the diametrically opposite point. At the points midway between these limits the planet meets the ecliptic as it passes through its nodes. (Copernicus makes the line of nodes pass through the center of the earth's orbit, not through the sun.) The cycle of changes in the planet's latitude, however, further depends upon the motion of the terrestrial observer, the latitude appearing to vary according to the planet's position in relation to the earth. This is partly an optical effect, due to variations in the distance between the two bodies; but Copernicus attributes it, in part, to a fluctuation in the inclination of the plane of the planet's orbit to the ecliptic. He likens it to the oscillation of the pole assumed in the theory of precession (Chapter IV, § 2, *supra*), and he supposes it to be regulated according to the following law (VI, 2):

Let ABCD (Fig. 36) be the earth's orbit with center E, and FGKL the planet's *mean* orbit inclined to the ecliptic, which it intersects at GEL. Let F be the northernmost and K the southernmost point on FGKL, and let G and L be the nodes. Let the *true* plane of the planet's orbit (at some given date) be OGPL, intersecting the mean orbit (and the ecliptic) at GEL. Then if

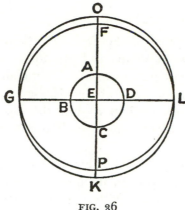

FIG. 36

the earth is at A and the planet at O, in opposition to the sun,
the inclination of its orbit is to have its maximum value (mean
value + maximum value of $\angle$OGF); if the earth is at B, 90°
distant in longitude from the planet, the inclination has its
mean value (O coincides with F); if the earth is at C, 180° from
the planet, the inclination has its minimum value (mean value
—maximum value of $\angle$OGF), whereas if the earth is at D, the
inclination has its mean value (O again coincides with F). The
planet may, of course, lie anywhere on its orbit, and is not
necessarily at O; but the general rule holds that the inclination
is a maximum when the planet is in opposition to the sun and
a minimum when it is in conjunction, the cycle of changes in
the inclination recurring in the planet's synodic period.

Copernicus deduces the mean inclinations of the orbits of the
several superior planets and the amplitudes of their oscillations
from observations of the planets' maximum angular departures
from the ecliptic at opposition and at conjunction (VI, 3; cf.
Alm., XIII, 3), and he shows how to calculate the apparent latitude of a superior planet with the earth at any given point on
its orbit (VI, 4).

(b) Case of An Inferior Planet

The closing chapters of the *De revolutionibus* treat of the motions in latitude of Venus and Mercury. In this portion of the work Copernicus can make but little claim to originality; he goes freely to Book XIII of the *Almagest* for observational data and for methods of disentangling the superposed effects of the several independent orbital oscillations postulated in his theory.

The orbit of each of the inferior planets is to be regarded as intersected by the ecliptic in its apse line; and the inclination of each is affected by two oscillations, the nature of which may be explained with the aid of Fig. 37 (VI, 2). Let ABCD be the

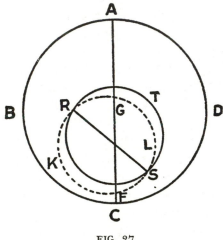

FIG. 37

earth's orbit, and let the dotted circle GKFL be the mean orbit of the planet (the eccentricity of which to ABCD is greatly exaggerated in the figure). The two planes intersect in the apse line FG. When the earth was at B or D, 90° from an apse, the inclination of the planet's orbit, deduced from observations of its latitude, was found to be less than when it was observed from A or C. This is accounted for by assuming an oscillation about the apse line FG causing the inclination of the planet's true orbit to be a maximum when the earth is at an apse, and a minimum when the earth is midway between the apses. When, however, the earth is at A and the planet at G, or the earth at C and the planet at F, the planet's latitude is not, in general, zero. Accordingly a second oscillation is assumed, called the *deviation*. The planet is supposed to travel on a circle, RTS, intersecting the plane GKFL at RS. The plane of this circle oscillates about RS, alternating between equal and opposite inclinations to GKFL. The planet is at R or S when the earth is midway between the apses, and at T, 90° from R and S, when the earth is at A, for RS swings round in such a manner that the orbital motion of the planet relative to T is the same as that of the earth relative to A. The *period* of this second oscillation is equal to the period in which the planet traverses the circle RTS relative to T, and therefore equal also to the period in which the earth completes the circuit ABCD. The *phase* of the oscillation about RS is so adjusted that the inclination of the circle RTS to the plane GKFL is a maximum when the planet is at T, which it is when the earth is at A. Hence, while the planet moves from T to R on the circle RTS, the inclination of this circle to the mean orbit diminishes from its maximum value to zero. Following out the cycle it becomes clear that the planet must lie always on the same side of the plane GKFL, touching it only in passing through R and S. Thus Venus is always north of GKFL and Mercury always south. The circle RTS is for Venus concentric but for Mercury nonconcentric with the orbit GKFL.

Copernicus devotes four chapters (VI, 5–8) to the evaluation of the numerical constants defining the motions of Venus and Mercury in latitude. We shall restrict ourselves to a brief outline of his procedure, suppressing all numerical details.

The first quantity to be examined (VI, 5) is the "declination," i.e., the latitude of the planet when the earth is midway between the apses of the planet's orbit and when, therefore, the "deviation," as defined above, vanishes. Ptolemy gives (*Alm.*, XIII, 3) values for the corresponding maximum north and south latitudes of Venus and Mercury when these planets are (*a*) nearest the earth and (*b*) farthest from the earth. From these values and the known distances of the planets at such times, Copernicus calculates the angles of inclination of the planes of the orbits to the ecliptic at the times of observation, and he shows how to calculate the latitude of each planet when it is at any given point on its orbit (with the earth still supposed to be midway between the apses), and also the amount by which this displacement in latitude alters the longitude of the planet measured along the plane of its orbit (cf. *Alm.*, XIII, 4).

The "obliquation" is next considered, i.e., the departure of an inferior planet from the ecliptic, observed when the earth lies in the planet's apse line (VI, 6, 7). The "deviation" is involved in this. The inclination of the orbital plane to the ecliptic, under these conditions, *unaffected by deviation,* is obtained from the *mean* of the latitudes of the planet, observed at its greatest elongations on the two sides of the sun. This value of the inclination is compared with that already found when the earth was midway between the apses, and the comparison gives both the mean inclination of the orbital plane and the amplitude of its oscillation about the apse line.

The "deviation" alone now remains to be determined (VI, 8). It was introduced to account for the fact that the maximum north latitudes of Venus, observed from the earth at an apse, exceed the maximum south latitudes, while the corresponding south latitudes of Mercury exceed the north latitudes. The mean values of the north and south latitudes (the "obliquation"), observed when the earth is at an apse, being known (VI, 6) for both planets, the residual northward displacement of

Venus and southward displacement of Mercury must be attributed to the deviation. This quantity can therefore be determined from a comparison of the mean and extreme values of the latitudes. When the earth is not at an apse, only a proportion of the maximum deviation has to be applied; and Copernicus shows (VI, 8) how to compute the correction appropriate to any given position of the earth in relation to the apses. This section of the *De revolutionibus*, which occupies the last few pages of the work, concludes with tables of planetary latitudes; these are similar in form to those of Ptolemy (VI, 8, 9; cf. *Alm.*, XIII, 5, 6).

The planetary theory of Copernicus represents not only his most sustained mathematical effort but also the most substantial evidence that he could advance of the scientific truth of his system. When the Copernican hypothesis was first formulated, in 1543, it was impossible to point to any celestial phenomenon which could not have been represented with equal accuracy by the Ptolemic theory. By adopting heliocentric coordinates, however, Copernicus was able to represent the familiar phenomena with greater economy of thought and to establish the mathematical necessity of certain relations long known to subsist among them but previously accepted as mere coincidences. It was on that account, and in that sense, that the Copernican theory could be regarded as more "true" than the Ptolemaic. It may be convenient to summarize here the chief mathematical advantages of the new planetary layout; we have already noted several of them in their appropriate contexts.

In the first place, by making the earth rotate on an axis and revolve in an orbit, Copernicus reduced by more than half the number of circular motions which Ptolemy had found it necessary to postulate. He also satisfied the sense of fitness which demands that the earth shall move rather than the vast fabric of the heavens. More particularly, he was able to dispose of the ancient dispute as to how Mercury and Venus are related to the

sun in the cosmic order (Chapter III, § 4, *supra*). By assigning
to these two planets sun-centered orbits interior to that of the
earth, he was able to rationalize what had appeared to early
astronomers as the mere coincidence that the centers of the sun
and of the respective epicycles of Mercury and Venus always lie
upon a straight line passing through the earth, these three
points each requiring the same period (one year) to perform a
circuit of the zodiac. At the same time Copernicus was able to
explain why Venus and Mercury (unlike the other planets and
the moon) must keep within restricted limits of angular distance from the sun. And by comparing these limits, he was able
to place the two planets in their correct order of increasing distance from the sun.

In addition, Copernicus was able to account for certain pecu-
liarities in the behavior of the three superior planets. It was
known that Mars appears brightest (and must therefore be nearest to the earth) when it rises at sunset, i.e., about the time of its
opposition to the sun. This had been explained by arbitrarily
supposing Mars to describe its principal epicycle in its synodic
period, the radius in the epicycle (CP in Fig. 3) always remaining parallel to the line joining earth and sun; and the cases of
Jupiter and Saturn were similar. But the reason for the phenomenon was obvious once it was admitted that the earth and
Mars describe about the sun, each in its own sidereal period,
orbits approximating to concentric circles. It was equally obvious, on the same supposition, why the retrogressions of all three
superior planets recur in their respective synodic periods, about
the times of opposition; and why a more distant planet shows
a smaller arc of retrogression. It was, in fact, the historic achievement of Copernicus to have related to a single cause those principal inequalities of the several planets which had always been
regarded as just so many unrelated phenomena to be separately
explained by the introduction of an *ad hoc* complication into
the economy of each planet. Copernicus, indeed, did not en-

tirely abolish such complications, for we have seen such factors as epicycles playing a prominent part in his system. But he initiated the process by which such arbitrary elements have been steadily eliminated from astronomical theories, and cosmic phenomena have been referred to ever fewer laws of ever increasing generality.

7

The Establishment of the
Copernican Theory

WITHIN ABOUT A CENTURY AND A HALF OF ITS FORMULATION BY Copernicus the heliocentric theory had gained almost universal acceptance in scientific circles. We shall now trace briefly the progress and development of the Copernican doctrine during the momentous period which opened with the publication of the *De revolutionibus* and culminated in the classic synthesis of Newton's *Principia*. Nothing like a complete account of the movement will be attempted here; we shall merely glance at the contributions of a few outstanding and representative participants in the great debate.

§ 1. THE WITTENBERG SCHOOL AND THE PRUTENIC TABLES

The climate of philosophical and religious opinion in Reformation Europe was unfavorable to the progress of the Copernican theory, which was opposed by both the Catholic and the Protestant parties. The geocentric cosmology formed an essential element of the scholastic philosophy entrenched in the universities—of Wittenberg no less than of Paris. The policy of Rome took some seventy years to crystallize, for it was in 1616 that the fundamental theses of the heliocentric theory first incurred formal ecclesiastical censure. On the other hand, the Protestant leaders, undeceived by Osiander's Preface, showed from the first a pronounced hostility to the new theory on the ground that it conflicted with the plain words of Scripture upon which, in the absence of any other historic authority, they took their stand.

Even before the publication of the *De revolutionibus* an account of its teachings had come to the ears of Martin Luther, moving him to testify against the "new astronomer who wants to prove that the earth goes round, and not the heavens, the sun, and the moon. . . . The fool will turn the whole science of astronomy upside down. But, as Holy Writ declares, it was the sun and not the earth that Joshua bade stand still." (Luther's *Tischreden,* ed. Walch, 1743, p. 2260.)

The heliocentric theory was assailed in more sober terms by Luther's fellow worker Philip Melanchthon, in his *Initia doctrinae physicae* (1549, with some sixteen later editions), a Christianized epitome of Aristotelian physics and Ptolemaic astronomy. Melanchthon summarizes the traditional arguments against the motion and eccentricity of the earth; these he corroborates with some of the texts most frequently quoted in support of the geocentric position. Some examples are: "In them hath he set a tabernacle for the sun, which is as a bridegroom coming out of his chamber, and rejoiceth as a strong man to run a race. His going forth is from the end of the heaven, and his circuit unto the ends of it: and there is nothing hid from the heat thereof." (*Ps.* 19: 4–6.) "Thou hast established the earth, and it abideth" (*Ps.* 119:90).

The efforts of the Reformers to check the spread of Copernican doctrines among their followers were, however, less effective than the corresponding measures of the Church of Rome later proved to be. For the Protestant Churches lacked both the long-standing and unique authority of the Papacy and the instruments of repression which were at its command; and it is noteworthy that several of the most prominent of the early Copernicans were drawn from the Protestant ranks.

Even in Wittenberg the *De revolutionibus* attracted the favorable notice of practical astronomers concerned with the construction and use of planetary tables rather than with the subtleties of natural philosophy. The heliocentric theory offered them a convention upon which the data of observation could be systematized, and future configurations predicted, as effectively as upon the Ptolemaic scheme but with greater mathematical simplicity.

Rheticus, the young professor from Wittenberg whose youthful enthusiasm cheered the last years of Copernicus, was, in all probability, the earliest of his disciples; and it was mainly through his enterprise that the *De revolutionibus* was published. However, the first important contribution after 1543 to the spread of Copernican ideas was the work of a colleague of Rheticus, Erasmus Reinhold (1511–53), who held the senior chair of mathematics at Wittenberg. Although compelled by the terms of his appointment to teach the Ptolemaic system, Reinhold prepared a revised and enlarged edition of the astronomical tables of Copernicus and published it in 1551 under the title of *Prutenicae tabulae coelestium motuum*. These tables followed, in the main, the rubrics laid down in Copernicus' book; but they were based upon an independently calculated set of constants and they aimed at greater precision and convenience. Although resting upon highly inadequate foundations, the Prutenic Tables, with occasional revisions, remained the standard ones of their kind for about eighty years. They represented some slight advance in accuracy over the Alfonsine Tables, and this, with their greater simplicity, led to their being generally adopted by practical astronomers, who were thereby disposed to regard more favorably the planetary theory associated with them. Reinhold deferred matters of theory to his *Commentary* on the *De revolutionibus,* which he did not live to print and which was formerly supposed to be lost; part of it (consisting of calculations relating to Books III–V) appears to be preserved in the Berlin State Library (E. Zinner: *Entstehung und Ausbreitung* etc., p. 513).

§ 2. SOME ENGLISH DISCIPLES

It is not surprising that some of the earliest literary evidences of the spread of Copernican ideas are to be found among scientific writers in England, where the Aristotelian philosophy had never superseded the Platonism of the Middle Ages to the same extent as it did in thirteenth-century France, and where greater freedom of thought prevailed.

There appeared in 1556 an almanac for the year 1557 (the first of a series) by one John Feild (or Field) (*c.* 1525–87), a Yorkshire country gentleman who taught mathematics for a time in London. Feild's almanac (*Ephemeris anni 1557 currentis juxta Copernici et Reinholdi canones . . . supputata,* London, 1556) was based upon the Prutenic Tables. It contains allusions to Copernicus, Rheticus, and Reinhold, stating, for example, "I have published for you this almanac for the year 1557, and in its preparation I have followed as authorities N. Copernicus and Erasmus Reinhold, whose writings are established and based upon true, sure, and authentic demonstrations." The work is prefaced with a letter from Dr. John Dee, the celebrated Elizabethan mathematician and occultist, who rebukes the makers of almanacs for paying no attention to the works of Copernicus "for the more than Herculean labours he has undergone in restoring astronomical science and in confirming it by the strongest rational proofs, of whose hypotheses this is not the place to treat," or to those of Rheticus and Reinhold "for the eager diligence they have shown in following in his footsteps." However, neither here nor in his later writings does Dee avow any definite opinion as to the physical truth of the Copernican theory.

In the same year (1556), but probably a few months later than Feild's *Ephemeris,* there appeared the *Castle of Knowledge* (London, 1556) of Robert Recorde (*c.* 1510–58), physician, Greek scholar, mathematician, and educational reformer. This vernacular primer of astronomy is in the form of a dialogue between a Master and a Scholar. In the fourth of the Treatises into which it is divided, the Master describes the fundamental phenomena from the common-sense point of view of a stationary observer surrounded by rotating stellar and planetary spheres; he then goes on to discuss whether the earth is at rest or in motion in words which show at least a just appreciation of the significance of the Copernican theory.

In November 1572 a "supernova," or temporary star of exceptional brilliance, suddenly appeared in the constellation Cassiopeia. This apparition set on foot investigations which precipitated the downfall of the old Aristotelian cosmology and so helped to create a situation more favorable to the acceptance of the Copernican theory. One of the astronomers whose curiosity it awakened was the Englishman Thomas Digges (c. 1545–95), a pupil of Dr. Dee. Digges was to play a leading part in propagating the heliocentric theory among his countrymen. His first concern was to fix the new star's position upon the celestial sphere and his next was to determine its distance from the earth. As the celestial spaces beyond the moon's sphere were traditionally conceived to be incapable of substantial change (see Chapter I, § 1, *supra*), there was a strong presumption that the new star would be found to lie on this side of the moon, in the earth's atmosphere.

Digges set himself to measure the optical effect known as the *diurnal parallax* of the star, i.e., its apparent displacement relative to the background of the permanent stars in the course of a night, which should have been the more pronounced the nearer the object was to the terrestrial observer. The moon's horizontal parallax was easily measurable by Digges with his crude apparatus; and from the shift suffered by the new star as between its upper and lower transits across the meridian (both observable) it would be possible to decide whether the object was nearer to us than the moon or more remote. Digges found the parallax imperceptible, and he concluded that it must be less than two minutes of arc (the limit of the resolving power of the eye); the inference was that the object must be located in the depths of the "immutable" heavens. However, in his book *Alae seu scalae mathematicae* (London, 1573), in which this result is recorded, Digges went further and took occasion to suggest that the changes in the star's brightness might furnish a means of confirming the truth of the Copernican theory. His argument was that, if the earth revolved round the sun, it must, in general, alternately approach and recede from any particular star, which might accordingly be expected to show annual fluctuations in brightness. He thought the gradual fading of the apparition in Cassiopeia might be due to the fact that the earth was just then withdrawing from that part of the heavens; after a few months it should begin to grow brighter again, and thereafter it should fluctuate with an annual periodicity. Digges did not explain why the other stars do not behave in this manner; and he must have been soon disappointed so far as the new one was concerned. In the dedication of his book, he promised to give a convincing demonstration of the truth of "the hitherto derided paradox of Copernicus concerning the motion of the earth." He actually began some *Commentaries upon the Revolutions of Copernicus;* but a succession of lawsuits in which he became involved left him no leisure to complete them before his death.

In the meantime, however, Thomas Digges had made a more important contribution to the propagation of Copernican ideas in England. In 1576 he brought out a new edition of *A Prognostication Everlasting,* a perpetual almanac first published under that title in 1556 by his father, Leonard Digges. The original work had included an outline of the Ptolemaic system of the world; Thomas Digges let this stand, but he added an appendix of his own (*A Perfit Description of the Caelestiall Orbes* etc.) consisting essentially of a free translation, into Elizabethan

English, of those passages from Book I of the *De revolutionibus* in which Copernicus gave a general account of his reformed system and answered the traditional objections to the hypothesis of the moving earth. The appendix is illustrated with a diagram of the heliocentric universe (Fig. 38), the first of its kind

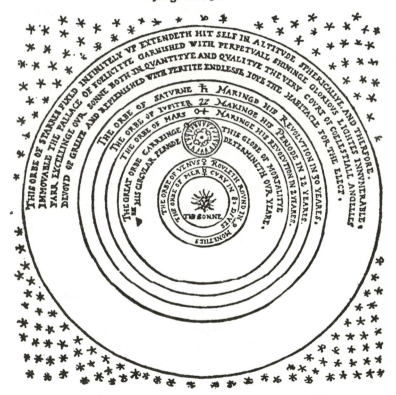

FIG. 38. The Copernican Universe of Thomas Digges (From T. Digges, *A Perfit Description of the Caelestiall Orbes*, 1576. Courtesy of the British Museum)

known to have been published in an English book. Both the diagram and the text contain notable additions to the original Copernican scheme. The earlier astronomers had supposed the fixed stars to be finite in number and to be distributed over the surface of a crystal sphere forming the boundary of space. But Digges wrote of the heavens as an infinite region through which an infinity of stars are scattered; his diagram shows the stars spreading out into space, and it bears the legend "This orbe of starres fixed infinitely up extendeth hit self in altitude sphericallye."

At the end of the sixteenth century, William Gilbert (*c.* 1540–1603) published his historic book *De magnete* (London, 1600). Gilbert was one of the founders of electrical science; and much of his book is taken up with the establishment of a correspondence between the magnetic properties of a spherical lodestone and those of the earth. In the closing pages of the work, the author urges the probability that the earth rotates about its magnetic poles. Gilbert considered it more credible that the earth should turn in twenty-four hours than that the rest of the universe, extending indefinitely (as he supposed) in all directions, should be carried round while the earth alone remained unaffected by the universal westward swirl. He points out, moreover, that the characteristic revolutions of the heavenly bodies are directed from west to east, and that they are completed in periods which, as we pass outward from the center, increase steadily from the 27 days of the lunar month to the 25,000 years of the precessional cycle of the stars. Is it likely, he asks, that beyond these there should be a body the revolution of which is contrary to that of all the others and requires only one day for its completion? If, however, a diurnal motion from west to east be assigned to the earth instead, these anomalies are obviated. On the other hand, Gilbert seems to have felt little interest in the hypothesis of the earth's annual revolution about the sun; and we find his hesitation between the Copernican and the geocentric planetary schemes more explicitly confessed in his posthumous book *De mundo nostro sublunari philosophia nova* (Amsterdam, 1651), where, however, Gilbert gives the Copernican explanation of the seasons and a diagram depicting the heliocentric universe with the stars scattered through space.

Another champion of the Copernican theory in its early days, less frequently cited than these English writers, was the Dutch mathematician Gemma Frisius. He expressed his unreserved support for the new astronomy in an introductory epistle contributed in 1555 to the *Ephemerides novae* (1556) of Ioannes Stadius (the passage was pointed out and reproduced in facsimile by Dr. Grant McColley in *Isis*, 1937, xxvi, 322).

§ 3. GIORDANO BRUNO AND THE UNBOUNDED UNIVERSE

The *Perfit Description* of Thomas Digges, being written in English, helped to make the Copernican system known among the author's countrymen; but it exerted correspondingly less influence upon the Continent. There the combination of the heliocentric theory with the conception of an unbounded universe was first found in the teachings of the Italian Giordano Bruno (1548–1600), a philosopher-poet rather than a man of science, who went considerably further than Copernicus in the novelty and boldness of his doctrines. Beginning his career as a Dominican monk, Bruno soon revolted against the orthodoxy of his time and forsook his Order to wander at large through Europe. He spent two formative years in Elizabethan London, where he may well have become acquainted with Thomas Digges or his writings, and where he published his two chief cosmological works, *La cena de le ceneri* and *De l'infinito universo et mondi* (both published in 1584); but his journeyings ended in the prison of the Roman Inquisition, whence he was led to the stake as an obstinate heretic.

The Greek atomists had taught that space extended without limit in every direction and that it contained infinitely many planetary systems, of which ours was but one. Plato and Aristotle threw the weight of their authority behind the rival conception of a unique and finite cosmos. As this view also appeared to agree with the Bible, it remained dominant throughout the Middle Ages, except among those who held that the infinite potentiality of God must find realization in an infinite work of creation. In the speculative ferment of the Renaissance, however, the doctrines of the atomists were revived, together with those of other submerged Greek schools. Now an infinite space can have no central point; and to anyone desiring to revive the atomists' cosmology, the heliocentric theory (which at least dethroned the earth from the central position) must have appeared a step in the right direction. It was probably on that account that Bruno passionately espoused and championed the Copernican doctrine as far as it went. For his own part, he went much further, propagating, in speech and writing, his conviction that the universe is infinite and eternal and that the stars are innumerable suns, each moving spontaneously through space accompanied by its train of inhabited planets. Bruno's tragic fate momentarily discouraged speculation along the lines associated with his teachings; but when interest was allowed to revive in the "plurality of worlds," it was the sun-centered planetary system which was adopted as the supposed unit of pattern endlessly repeated throughout space.

The extremism of Bruno's ideas may well have helped to awaken the Church of Rome to the heretical implications of the new astronomy and thus at length to bring the Copernican doctrine under the ban of the Inquisition. The old conception of the earth as resting at the center of a finite universe fitly symbolized the centrality of man in an order of nature created only to minister to his needs and destiny. The Copernicans, who would deprive man of his privileged place in the cosmos, had to face the same kind of opposition as the nineteenth-century evolutionists, who denied man's unique status in the world of life.

However, a still more drastic development was implicit in the Copernican doctrine. Aristotle had argued that the universe must be finite, for it rotated once in a day, and any part infinitely distant from the center would never get round in that time unless it traveled at an infinite speed, and he declared this to be impossible (*De caelo*, I, 5). Copernicus did not debate the finitude of the universe, but by transferring the diurnal rotation from the heavens to the earth he destroyed the force of Aristotle's argument. There was nothing now to check the growing conviction that space extends in all directions without assignable limits. Thomas Digges, although he accepted the infinity of the universe, maintained that the solar system occupied the central position in a boundless "orbe of starres" constituting "the gloriouse court of the great God." But Bruno logically denied that an infinite space could have any center or boundary; and he reduced the status of the sun to that of one star among the rest. By the time he had completed the disintegration of the Aristotelian cosmos, there remained nothing to distinguish this earth, the scene of the drama of Redemption, from its myriad sister-globes, and no empyreal heaven whence God could look down upon His creatures. It was, significantly, the generation following Bruno which saw the Church of Rome extend its repressive control from purely philosophical and theological doctrines to include scientific opinions as well. This new policy found expression in an anti-Copernican drive of which Galileo was to be the first notable victim.

§ 4. TYCHO BRAHE AND THE REVIVAL OF OBSERVATION

The Prutenic Tables (see § 1, *supra*), which commended the Copernican theory to sixteenth-century astronomers, were based upon recorded celestial observations, at best made with crude instruments and at worst spurious or corrupted by copyists' errors. A series of planetary observations, more refined and systematic than any previously made and extending over many years, was the indispensable and timely contribution of the astronomer Tycho Brahe (1546–1601) to the establishment of the Copernican theory. Born of a Danish noble family, Tycho Brahe spent his youth in foreign travel and study, devoting to the pursuit of astronomy the years that should have been dedicated to law and rhetoric. Returning to Denmark, he was endowed by the King, Frederick II, with a palatial observatory where he labored from 1576 to 1597. Tycho's career as an observer was eventually cut short by the death of his royal patron. In 1598 he accepted an invitation from the German Emperor, Rudolph II, to settle in Prague, where he sought to resume his former activities; but in 1601 he succumbed to a sudden illness.

The apparition of the "new star" of 1572 (see § 2, *supra*) afforded Tycho Brahe his earliest opportunity of contributing to the reform of cosmology. Like Digges, he established that the object exhibited no diurnal parallax and must therefore occupy a place in the stellar domain. In like manner Tycho was able to assign the comet of 1577 (and other such visitants in later years) to the superlunary realm. He had at first shared the common opinion that the heavens were filled with material spheres carrying the planets round in their courses; but the spectacle of comets traveling freely through interplanetary space convinced him that such spheres, which even Copernicus had not explicitly renounced, could have no real existence (*Opera*, ed. Dreyer, vii, 130). For they evidently presented no obstacle to the passage of comets; moreover, these bodies themselves must presumably move through the heavens without the support of carrying-spheres. The clearing away of such machinery was all in favor of the Copernican theory. For if the planets moved without any visible external cause, why not the earth also?

Tycho Brahe always wrote of Copernicus with the highest admiration; he commended him particularly for rejecting the Ptolemaic practice of reckoning the circular motion of a planet as uniform about a point other than the center of the circle. But he felt compelled to reject the hypothesis of the moving earth on physical and Scriptural grounds: "When [Copernicus] declares that the gross, slothful body of the earth, unfit for motion, is driven along with no slacker course (nay, rather, with a threefold motion) than the celestial luminaries, he is contradicted not only by the principles of physics but also by the authority of the Holy Scriptures, which repeatedly confirm the stability of the earth" (*Opera*, iv, 156). An argument of greater validity which he directed against the annual revolution of the earth was the seeming absence of any corresponding parallactic displacement in the apparent positions of the stars. Tycho accordingly proposed an alternative planetary hypothesis of his own (mathematically equivalent in principle to the Copernican scheme), according to which the earth was stationary at the center of the universe; the moon revolved about it in a month and the sun revolved about it in a year; and the five planets revolved, each in its own characteristic period, about the sun, the superior planets embracing the earth and the moon within their orbits. All these bodies also shared with the fixed stars in the daily revolution about the earth.

The arguments which Tycho Brahe marshaled against the Copernican theory and in favor of his own were set forth chiefly in letters to his friend Christoph Rothmann, an early German Copernican, scarcely remembered except through his interesting correspondence with the Danish astronomer (*Epistolae astronomicae*, Uraniborg, 1596; *Opera*, vi). Employing an argument which was to become familiar in the later stages of the controversy, Tycho demanded of Rothmann what would happen if two equal projectiles were fired under precisely similar conditions, one toward the east and the other toward the west.

All experience proves that they will travel equal distances over the earth's surface. But, he holds, if the earth were rotating from west to east, the range of the projectile fired toward the west would be greater than the range of that fired toward the east; in the former case the projectile and the surface of the earth are moving in opposite directions and the apparent range is the *sum* of their respective displacements during the time of flight, whereas in the latter case they are moving in the same direction and the apparent range must be the *difference* of the two displacements (*Opera*, vi, 219). Any natural tendency which the projectile might have to move eastward with the earth would, Tycho thought, be suspended during the time of flight, just as (according to the dynamical ideas which still prevailed) the tendency of a heavy body to move toward the center of the earth was temporarily suspended when the body was set in violent motion (e.g., by being thrown) and reasserted itself only when that motion had decayed.

However, Tycho Brahe's opposition to the Copernican theory is far outweighed in significance by the services which he rendered to the new astronomy as one of the greatest observers of all time. He improved upon the design and construction of the types of instruments employed in his day for measuring the positions of the sun, moon, and planets in relation to the background of stars and to the standard celestial circles; and the long series of planetary observations which he was thus able to leave to his successors was all the more valuable because accompanied by an entirely new star catalogue, which completely superseded that of Copernicus.

Shortly before his death, Tycho Brahe had added to his little band of assistants in Prague a young and comparatively unknown German astronomer, Johannes Kepler; and it is with his name that the next notable stage in the establishment of the Copernican theory is associated.

§ 5. JOHANNES KEPLER AND THE TRANSITION TO DYNAMICAL ASTRONOMY

Of all the various contributions to the triumph of the heliocentric theory, that of Kepler was so characteristic of the personality and outlook of its author that it was the least likely to have been made by anyone else. Had he not lived, or had the observations of Tycho Brahe fallen into other hands, the progress of astronomy and of physical science generally might have been held up for generations. It is possible to regard Copernicus as the last of the ancient astronomers rather than as the first of the modern ones; he had but given a new turn to an age-long quest. Kepler, however, in closing the canon of classical planetary theory, supplemented the geometrical conventions with an auxiliary physical hypothesis. He thus inaugurated the dynamical regime in astronomy under which all rational opposition to the heliocentric theory was soon to disappear. Moreover, his writings represent a transitional phase between the medieval conception of the closed universe, sharply divided into orderly heavens and disorderly earth, and the indefinitely extended cosmos of Descartes or Newton, throughout which a uniform system of natural law everywhere prevails.

Born in 1571 in Württemberg, Kepler was won over to the heliocentric theory as an undergraduate at Tübingen through the teachings of his Copernican professor, Michael Mästlin (c. 1550–1631), whose observations of the new star of 1572, establishing the insensibility of its diurnal parallax, had been particularly commended by Tycho Brahe. Mästlin had convinced himself that the course through the heavens followed by the comet of 1577 agreed with one of the component motions which Copernicus had assigned to Venus and was to be explained by supposing the object to have become attached somehow to one of that planet's "spheres" (*Observatio et demonstratio cometae aetherei*, Tübingen, 1578.)

Kepler's earliest cosmological work, *Mysterium cosmographicum* (1596), contains several of the germinal ideas of which his subsequent achievements were the ripe fruit. Influenced, like Copernicus, by the neo-Pythagorean doctrines of the age, Kepler formed the conviction that such natural specifications as the number of the planets and the sizes and periods of their orbits were not fortuitous but represented the aesthetic choice of the Creator.

In the first chapter of his *Mysterium,* Kepler explains the considerations which led him to adopt the heliocentric theory. He proves that the Ptolemaic epicycles of Mars, Jupiter, and Saturn represent the result of transferring to these planets an annual motion which actually belongs to the earth, and that these epicycles subtend at the eye of the terrestrial observer the same angles as would the earth's orbit if it were viewed from the distances of the respective planets. Furthermore, the heliocentric theory alone can explain, or render self-evident, the otherwise arbitrary fact that each of the superior planets describes its principal Ptolemaic epicycle in its synodic period, approaching nearest to the earth when in opposition to the sun. The main purpose of the book, however, was to establish that the relative distances of the six planets from the sun had been determined by the Creator according to certain geometrical relations among the five regular solids of the Greeks. Kepler found that the planets would fit better into his geometrical scheme if he measured their distances not from the center of the earth's orbit (the "mean sun") as Copernicus had done, but from the actual sun. It may be recalled that each of the planets revolves round the sun in an orbital plane of its own which is inclined at a moderate angle to the ecliptic and which intersects the latter in a "line of nodes." In the original planetary scheme of Copernicus that line was supposed to pass through the center of the earth's orbit, not through the true sun. Kepler, however, preferred to regard each planet's orbital plane and its line of nodes as pass·

ing through the sun itself, from which the planetary distances were to be measured. This scheme not only gave better agreement with the scheme of the regular solids but had the further advantage of treating the earth just like any other planet, whereas Copernicus had assigned to the earth a privileged status by making the center of its orbit the point of intersection of the lines of nodes of the other planetary orbits. These lines were now to intersect in the sun, which was thus to occupy the cardinal position in the planetary system; and the way was opened for explaining the motions of the planets as due in some way to the agency of the sun. Thus the change of origin acquired a deeper, physical significance.

Kepler wished to test his hypothesis against the refined planetary observations of Tycho Brahe, whom he joined in Prague early in 1600. A short period of uneasy collaboration between the two men ended with the death of the Danish astronomer, whose observations Kepler eventually obtained. Tycho had assigned to his young assistant the task of establishing the theory of Mars, a fortunate choice, since Mars was the only planet the departure of which from the ideal circular orbit could at that period have been readily detected. How Kepler carried through this task is vividly narrated in his epoch-making *Astronomia nova* (1609).

Of the five parts into which that work is divided, the first institutes a comparison of the Ptolemaic, the Copernican, and the Tychonic representations of the planetary inequalities; and it stresses the advantage of referring the planet's motion to the true instead of the mean sun even in the older systems with which readers would be chiefly familiar. Kepler urged this reform upon Tycho Brahe as soon as they began working together. One immediate consequence of the change was that the orbital plane of Mars and eventually those of the other planets were found to be inclined at constant angles to the ecliptic, and not subject to the fictitious oscillations about their lines of nodes which take up most of Book VI of the *De revolutionibus*. On the other hand, whereas Copernicus had represented the motion of a planet by means of an eccentric and an epicycle, both uniformly described about their centers, Kepler abandoned the epicycle and reckoned the planet's angular velocity as uniform about an equant point H (*punctum aequans*—center of uniform angular velocity), coinciding neither with the true sun, S, nor with the center, C, of the planet's circular orbit (Fig. 39). This

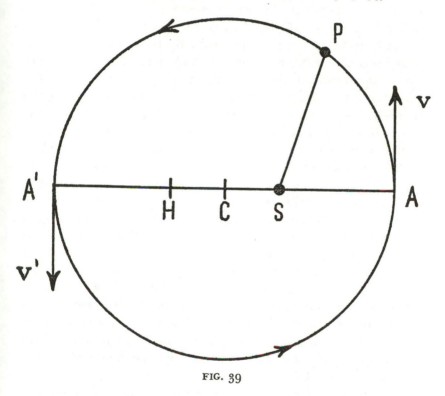

FIG. 39

was a return to the practice of Ptolemy (objectionable to Aristotelian physics, as well as to Copernicus, who insisted that a planet should revolve uniformly about the *center* of its circle).

Kepler did not at first assume with Ptolemy that CS = CH; but he soon found that this "bisection of the eccentricity" gave the best results, under the restriction to a circular orbit. One reason why Kepler reintroduced the equant was the very reason why Copernicus had given it up. It implied that the planet traversed its orbit at a non-uniform speed, faster when it was nearer the sun and slower when more remote; and this, once again, was in accordance with Kepler's general view that an influence emanating from the sun kept the planet moving.

The second part of the *Astronomia nova* describes Kepler's attempts to fit such a circular orbit, by trial and error, to selected sets of four observations of Mars made when the planet was in opposition to the true sun. At such times, the sun, the earth, and Mars lie in a straight line, and the direction of the planet is the same whether viewed from the earth or from the sun. Thus uncertainties connected with the unknown motion of the earth do not affect the determination of the planet's place in its orbit. Kepler, however, could not rest satisfied even with the best-fitting orbit (correct to 8 minutes of arc in longitude), obtained after seventy trial applications of this method. Accordingly (as is related in Part III), he made a fresh attack upon the problem by way of an investigation of the *earth's* annual motion. By observing Mars at various dates when the planet was at the same point on its orbit, Kepler was able to calculate the corresponding values of the earth's longitude and distance from the sun, and hence to construct the earth's orbit and to determine its elements. The results answered to his expectation that the earth, like any other planet, would be found to have an equant equidistant with the sun from the center of the annual orbit (CS = CH).

Kepler now began to introduce physical assumptions in order to explain why a planet should not describe its orbit at a constant speed. The periods of revolution of the successive planets increase as we go outward from the sun, and this cannot be accounted for entirely by the greater lengths of the outer orbits. Hence Kepler had long suspected that whatever agency drove the planets along must grow weaker with increasing distance from the sun. He came to conceive this agency as some kind of solar emanation, carried round with the supposed axial rotation of the sun and bearing the planets along with it: "upon the more remote bodies the force is in some degree weakened owing to the distance and the resulting attenuation of the virtue" (*Mysterium,* cap. xx; *Opera,* ed. Frisch, i, 174). Influenced by his reading of Gilbert, Kepler regarded the action of the sun upon a planet as at least analogous to that of a magnet upon iron. He conceived the solar virtue as restricted to the plane of the ecliptic; hence it seemed reasonable to assume that its intensity would vary inversely as the simple distance from the source. Now assuming the velocity of a planet to be proportional to the force applied (the traditional mechanical doctrine), and this force to vary inversely as the simple distance from the sun, then the velocity, too, must vary inversely as the distance. Kepler, we saw, had been driven to conclude that the earth's orbit had an equant H where $CH = CS$ (Fig. 39). For a body describing such an orbit, the velocities v and v′ at the apses A and A′, are readily shown to vary inversely as the distances from the sun $(v/v′ = HA/HA′ = SA′/SA)$.

Kepler tentatively assumed this law to hold of *every part* of the earth's orbit; and he had now to prove that it led to results agreeing as well with observation as the hypothesis of uniform motion about the equant H. If the speed of the earth varied inversely as its distance from the sun, the time required for it to traverse a given small element of its orbit must vary directly as this distance. Hence, as a rough test of the law, Kepler divided the earth's circular orbit into 360 equal arcs, calculated the distances from the sun to each point of division (taking the radius CA as unit), and then tried to find whether the sum of these distances, say, between A and P was proportional to the time

taken by the earth in going from A to P. But the calculation was
tedious, and the results were only approximate. Kepler's thoughts
turned to the method of splitting a circle into an infinity of
triangles, by which Archimedes had found the ratio of the circumference of the circle to its diameter: "As I recognized that
there was an infinity of points on the eccentric, and, corresponding to these, an infinity of distances [from the sun to the earth],
the thought occurred to me that all these distances were contained in the surface of this circle" (cap. xl). Hence he adopted
the convenient working rule that the time required for the
earth to travel from A to P was proportional to the *area* ASP
described by the radius vector SP, the total period of revolution
corresponding to the entire area of the circle. Kepler was aware
that the three assumptions (*a*) that the angular velocity is constant about H, (*b*) that the sum of the distances SP is proportional to the time, and (*c*) that SP sweeps out equal areas in
equal times do not give identical representations of the planet's
motion. He regarded (*c*) as a convenient if inexact substitute for
(*b*). For the earth, the difference proved to be inconsiderable,
because the eccentricity of its orbit is so small; and any of these
three assumptions sufficiently represented the facts.

Having thus investigated the motion of the earth and established the exact effect it must have upon the apparent motion of
Mars, Kepler returned (in Part IV) to the consideration of the
latter planet and sought to construct its orbit by a method corresponding to that already adopted for the earth. But he found
that, if the radius of the eccentric circle were chosen to fit the
planet's distances from the sun when near the apses of its orbit,
then, at other points, the planet lay not on but within this circle;
this showed that the orbit was some sort of oval figure, or, as he
eventually proved sometime in 1605, an *ellipse,* having the sun
in one focus and described according to the above-mentioned
"area law" (cap. lix; *Opera,* iii, 401). Part V of the *Astronomia
nova* deals with the motions of the planets in latitude. Kepler's

third law of planetary motion, that the squares of the periods of revolution of the planets are as the cubes of their mean distances from the sun (the semi-axes major of their elliptic orbits) is to be found in the midst of the number-mysticism of his *Harmonia mundi* of 1619 (Lib. V, cap. iii; *Opera*, v, 279).

Kepler attempted to force his third law into relation with his physical hypotheses in another of his later works, *Epitome astronomiae Copernicanae* (1618–21). The *Epitome* was intended to give Kepler's views a wider currency than the *Astronomia nova*, which appeared in a small edition and which was too difficult to understand and too much in advance of contemporary ideas to exert much influence at the time. It is a textbook of general astronomy in the form of a catechism, emphasizing the gain in mathematical simplicity which would result from the adoption of the Copernican doctrine. From its great store of arguments for the various motions of the earth we select one referring to the daily rotation. The polar axis, about which that rotation (whether of the earth or the heavens) takes place, is fixed relative to the earth (as Kepler supposed), but it moves about in relation to the sphere of stars (thus giving rise to the precession). Hence the axis and the rotation about it belong to the earth more probably than to the stars. In the *Epitome* Kepler presents a more systematic and mechanistic account of the causation of a planet's motion. He compares the sun to a magnet having one pole at its center and the opposite kind of polarity distributed over its surface. The body of a planet is made up of parallel magnetic fibers, so that opposite faces of a planet correspond to opposite poles of a magnet, the one attracted and the other repelled by the sun. Thus, as the planet revolves in its orbit, presenting each of these two faces to the sun in turn, it alternately approaches and recedes from the luminary, and so the characteristic properties of the elliptic path receive a physical explanation.

In Kepler's *Epitome*, as in his *Mysterium* and his *De nova stella* (1606), the sun and its train of planets are assigned a unique position in the universe at the center of a vast starless sphere of space surrounded by stars; as to whether these lie upon the surface of a sphere or extend outward from it, perhaps to infinity, "astronomy makes no pronouncement" (*Opera,* vi, 138). Kepler's last contribution to the reformation of astronomy was his great set of planetary tables, *Tabulae Rudolphinae,* published in 1627; he died in 1630. These quickly superseded the tables in current use, and they held their own for nearly a century.

§ 6. GALILEO AND THE ATTACK ON THE TRADITIONAL COSMOLOGY

Kepler had originally embraced the Copernican hypothesis because it bound together in mathematical necessity numerous planetary phenomena which had previously appeared arbitrary and disconnected. He was confirmed in the doctrine when it afforded him the clue to his classic laws of planetary motion, which could not have been discovered or expressed in readily intelligible form except on the assumption that the earth and planets revolved round the sun. However, the Pythagorean and Platonist ideals which inspired Copernicus and Kepler were repugnant to those who still adhered strictly to the natural philosophy of Aristotle. And when Kepler himself attempted to furnish a physical explanation of why the planets moved as they did, his *ad hoc* magnetic hypothesis was not likely to convince anyone who had not already accepted the heliocentric theory on other grounds. The need of the age was for a new physico-mechanical synthesis which should make the Copernican doctrine seem as natural and inevitable as the geocentric structure of the universe had appeared to the disciples of Aristotle.

The foundations of such a synthesis were laid by Kepler's great contemporary, Galileo Galilei (1564–1642), who contributed to the triumph of the new cosmology along several different lines. In the first place, Galileo's researches in mechanics tended to neutralize the traditional arguments against the motion of the earth and to set the planetary problem in a new light. Secondly, with the aid of telescopes of his own device and construction, he discovered a number of celestial phenomena, some of which told strongly in favor of the Copernican theory. Lastly, Galileo stood out conspicuously as a champion and a brilliant exponent of the heliocentric doctrine. His critical faculties developed only gradually; but the studies of his early manhood in the works of Archimedes taught him that technique of combining idealized experiment with mathematical deduction which he did much to make the established procedure of physics. Entering upon a professional career in 1589, Galileo successively held university teaching posts in his native Pisa and in Padua; thereafter he became mathematician to the Grand Duke of Tuscany and so continued during the period of his conflict with the ecclesiastical authorities.

Galileo was formerly represented as having suddenly substituted modern mechanics for traditional doctrines unchallenged since their formulation by Aristotle. More recent researches have revealed the limitations imposed upon his thought by the scholastic training which he never entirely outgrew, and his debt to critical-minded predecessors, notably to his older contemporary Giambattista Benedetti (1530–90). A Venetian by birth, Benedetti figured prominently in a sixteenth-century revolt against the views of Aristotle as to how the speed of a freely falling body is determined and how the motion of a projectile is maintained. In his "Divers Speculations," in 1585, Benedetti had also expressed cautious approval of the Copernican doctrine.

Some of the speculations which Galileo committed to writing as a young teacher at Pisa already suggest an interest in the bearing of general mechanical principles upon the Copernican theory, although without explicit reference to the latter (*Opere*, ed. Favaro, i, 299, 304 ff., 373). He considers, for example, a sphere of elementary matter of which the center of gravity describes a circle about the center of the universe; and he asks whether such motion will continue of itself forever—whether, in fact, it is *natural* (like motion toward the center, maintaining itself) or *violent* (like motion away from the center, requiring external compulsion). He links the question with the problem of the motion of a ball rolling on an inclined plane (which he already had under consideration): a body revolving round the center of the universe corresponds to a ball rolling along an indefinitely extended smooth surface the inclination of which to the horizontal is made vanishingly small. Such a motion would indeed maintain itself forever; but it did not escape Galileo that the horizontal surface traversed by the rolling ball must be not a plane but a sphere concentric with the earth; and the motion continued indefinitely thereupon must be a circular motion about the earth's center. This consideration may well underlie Galileo's subsequent assumption that the planets naturally revolve in circles about the sun; and it may explain his indifference to Kepler's attempts to account for such revolution on physical principles. It would seem that Galileo first embraced the heliocentric doctrine in youth or early manhood; in a letter written in 1597, when he was thirty-three, he told Kepler that he had taken this step "many years ago." However, he had nearly reached his fiftieth year before he openly professed his Copernican faith while announcing his telescopic discoveries.

Galileo was not the first inventor of the telescope nor the only man of his day to utilize it for astronomical purposes. His achievements were closely paralleled by Thomas Harriot in England. However, the epoch-making results which followed Galileo's use of the instrument from 1609 go far to justify the traditional estimate of him as the founder of telescopic astronomy.

Many of the celestial discoveries announced in his *Sidereus nuncius* (1610) and in later tracts and letters suggested significant analogies or helped to break down illusory distinctions between the earth and the celestial bodies.

Turning his telescope to the moon, Galileo distinguished irregularities in this supposedly "perfect" sphere comparable to those upon the surface of the earth. It was in his account of these lunar observations that he first declared that he could furnish a hundred proofs that the earth revolves about the sun.

Observing next the planet Jupiter, in the early days of 1610, he discovered its four principal satellites. This revelation had a particularly important bearing upon the Copernican controversy. For the satellites of Jupiter revolved unmistakably about a body other than the earth, so that in future it would be impossible to regard the earth as the unique material center of all celestial motions. Moreover, the association of such satellites with an admittedly moving planet was calculated to demolish the arguments of those who refused to admit that the moon, while revolving about the earth, could at the same time accompany that body in an annual revolution about the sun, so describing a complicated orbit without any visible mechanism to constrain it.

Continuing his observations from Padua, Galileo detected mysterious appendages to the planet Saturn; for some years he followed their transformations (and temporary disappearance), but the quality of his instrument did not permit him to recognize in them a ring encircling the planet.

It had always been maintained by the older astronomers that, if Mercury and Venus revolved about the sun, as was required by the Copernican theory, they must either be self-luminous or translucent to the sun's beams. Otherwise they would not appear circular but would show phases analogous to those of the moon, owing to the varying degrees of obliquity with which their illuminated hemispheres would be presented to our view. They should also show occasionally as spots upon the solar disc. Galileo now established that Venus does, in fact, exhibit phases, proving its revolution about the sun (and not about the earth) and that it must be a dark body which, like the earth, receives its light from the sun. Thus Venus and the earth were both dark bodies; and as Venus revolved about the sun, there was a presumption that the earth might do so too.

Turning his attention next to the Milky Way, Galileo noticed that his telescope, which revealed an infinite multitude of stars, did not increase their apparent size as it did that of a body such as the moon. He thought that this was because the stars, being "fringed with sparkling rays," looked larger to the naked eye than they otherwise would—an effect we now call *irradiation*. He thus reduced the current estimates of the apparent size of the stars, so that it was no longer so acute a problem why the stars that looked so large were nevertheless too far away to show an appreciable annual parallax such as was to be expected according to the Copernican theory.

Galileo's relation to the discovery of yet another celestial "imperfection," namely, sunspots, is not quite so clear, although it is probable that he was the earliest to observe the spots through the telescope.

The book, published in 1613, in which Galileo described his work on sunspots was of a thoroughly Copernican tone; and about this time Galileo was drawn into the debate as to whether belief in the motion of the earth was contrary to the teachings of Scripture. In December 1613 he wrote a letter to his friend Castelli (and later, another to the Dowager Grand Duchess of Tuscany) in which he put forward the view that the Bible gives to men infallible guidance concerning the way of salvation, but it is not concerned to teach them astronomy, or anything else that they can find out for themselves by the exercise of their senses and reasoning powers. Moreover, he held that the words of Scripture were suited to the understanding of those for whom they were originally written; where they deal with scientific matters they must not be taken too literally; rather, they must be interpreted in the light of experimentally established facts. The letter to Castelli was made the excuse for denouncing Galileo to the Inquisition as a heretic in 1615. As a result of a visit to Rome in that year, Galileo cleared himself of this charge. However, early in 1616, the Qualifiers of the Holy Office were bidden to examine and pronounce upon the fundamental theses of the Copernican theory, namely, (*a*) "That the sun is the center of the universe and is wholly stationary," and (*b*) "That the earth is not the center of the universe, and is not stationary, but moves bodily and also with a diurnal motion." They declared (on February 24, 1616) that the thesis (*a*) was "foolish and absurd in philosophy, and formally heretical, inasmuch as it expressly contradicts the teachings of many passages of Holy Scripture, according to the sense of the words and to the common interpretation of the Holy Fathers and of learned theologians." The thesis (*b*) incurred the same censure as regards philosophy, and, in respect of theology, it was denounced as being at least an erroneous belief (*Opere,* xix, 321). At the same time Galileo was formally warned of the error of these doctrines, and a few days later (March 5, 1616) Copernicus' work *De revolutionibus* was placed on the Catholic *Index of Prohibited Books* "until it should be corrected."

The Edict of 1616 also suspended a *Commentary upon Job* (Toledo, 1584) by one Didacus à Stunica of Salamanca. He would appear to have been the first to interpret a Bible text as supporting the Copernican theory, and he had on that account been cited by Galileo in his letter to the Grand Duchess of Tuscany. It was the text "Which shaketh the earth out of her place, and the pillars thereof tremble" (*Job* 9: 6), which Didacus proposed to expound by reference to "the Opinion of the Pythagoreans, who hold the Earth to be moved of its own Nature. . . . But in this our Age, Copernicus doth demonstrate the courses of the planets to be according to this Opinion. Nor is it to be doubted but that the Planets Places may be more exactly and certainly assigned by his Doctrine, than by Ptolemies Great Almagest or Systeme, or the Opinions of any others." (For an English translation of the relevant pages of the *Commentary* see T. Salusbury, *Mathematical Collections and Translations,* 1661, i, 468 ff; and for a facsimile of Salusbury's translation, introduced by Dr. Grant McColley, see *Annals of Science,* 1937, ii, 179 ff.)

The Edict of 1616 formally debarred Catholics from taking part in the development of Copernican astronomy for about a century. Galileo pressed on, however, toward the completion of his great book on the "two chief systems of the world" (*Dialogo ... sopra i due Massimi Sistemi del Mondo Tolemaico, e Copernicano,* Florence, 1632). In this work he placed before his countrymen, in their native Italian, the claim of the new cosmology to be consistent with sound physical principles, and he marshaled the evidence which his own discoveries had brought to its support.

The book is in the form of a discussion, carried on among three friends, Salviati, Sagredo, and Simplicio, who meet on four successive days to debate the arguments for and against the Ptolemaic and the Copernican systems of the universe. Salviati and Sagredo bore the names of deceased friends of Galileo, and Simplicio was called after Simplicius, the great commentator on Aristotle. Salviati puts the case for the Copernican theory and is really the spokesman of Galileo's own opinions. Sagredo is an "inquiring layman" who draws Salviati out. Simplicio represents the conservative attitude which held to the traditional physical doctrines and to the geocentric theory.

The discussions of the first day are concerned mainly with the question whether the earth is stationary and occupies the center of the universe, or not. Salviati contends that, when once the world had been reduced from chaos to order, there was no reason why the terrestrial elements should move *naturally* in straight lines, since such motion would take them away from their proper places toward points which they could never reach. The natural motion should rather be a circular one, which carries bodies back to their starting points. The argument that a body detached by violence from the whole of which it is a part returns thither by a straight line as the shortest path does not imply that the whole earth, if removed forcibly from its place, would return to it again. For heavy bodies fall toward the center of the earth because it *is* the center of the earth, and not because it is the center of the universe. As for the alleged distinction between the corruptible earth and the incorruptible heavens, we cannot trust the evidence of our senses in this matter, since, even if changes took place in the heavenly bodies upon the same scale as the terrestrial changes occurring around us, we should not be able to detect them. In any case, direct evidence of the mutability of the heavens is afforded by the occurrence of sunspots, comets, and new stars.

The second day is devoted mainly to discussion of whether the earth has a daily rotation on its axis, producing the apparent rotation common to all the rest of the universe. Salviati, making use of some of Gilbert's arguments, finds it much easier to believe that the earth turns round once a day on its axis than that the incomparably greater celestial sphere does so. A motion of the sphere of stars from east to west would break the general rules that the heavenly bodies normally move from west to east and that, as we pass outward from the moon, the periods of revolution of the successive bodies steadily increase. According to the Ptolemaic hypothesis, precession must cause the distance of each star from the pole, and hence the extent of its diurnal circuit, to vary with lapse of time. Salviati further points out that, according to the ancients, while the stars and planets are whirled round from east to west, "the sole little Globe of the Earth pertinaciously stands still, and unmoved against such an impulse . . . nor can I see how the Earth, a pendent body, and equilibrated upon its centre, exposed indifferently to either motion or rest, and environed with a liquid ambient, should not yield also as the rest, and be carried about. But we find none of these obstacles in making the Earth to move; a small body, and insensible, compared to the Universe, and therefore unable to offer it any violence." (*Two Chief Systems,* Salusbury's translation, 103.)

Simplicio, taking up the line of argument pursued by Tycho Brahe, maintains that, were the earth in diurnal rotation, falling bodies would not drop perpendicularly but would be deflected westward, just as (he alleges) a stone falling from the masthead of a moving ship is deflected toward the stern; the range of a gun would be greater westward than eastward; birds and clouds would be carried westward by perpetual east winds, and objects near the equator would be hurled off into space. Salviati replies that the argument from falling bodies begs the question. For, supposing the earth to rotate, a falling body must have a motion compounded of (*a*) a perpendicular motion toward the center of the earth, admitted by both disputants, and (*b*) the circular motion about the center of the earth which the body must possess in virtue of its being a part of the rotating earth. Now the motion (*b*) is the same as that of the earth's surface and therefore cannot be detected; hence a body dropped from a tower must appear, in any case, to fall perpendicularly, striking the ground at the foot of the tower, and no conclusion can therefore be drawn from the alleged absence of any westward deflection. Moreover, Simplicio's example of the behavior of a stone let fall from the masthead of a moving ship is based on mere hearsay: "whoever shall examine the same . . . shall see the stone fall at all times in the same place of the Ship, whether it stand still, or move with any whatsoever velocity. So that the same holding true in the Earth as in the Ship, one cannot from the stone's falling perpendicularly at the foot of the Tower, conclude anything touching the motion or rest of the Earth." (*Ibid.,* 126.) Birds and clouds are carried along by the air, which shares in the circulation of the earth; and the centrifugal tendency of objects on our rotating globe is easily controlled by gravity.

The third day is devoted to arguments for and against the doctrine of the earth's annual revolution. Since the planets appear to move about the sun rather than about the earth, being nearest to us when in opposition, it is more reasonable to put the sun in the center of the universe: "rest seemeth with so much more reason to belong to the said Sun, than to the Earth, in as much as in a moveable Sphere, it is more reasonable that the centre stand still, than any other place remote from the said centre" (*ibid.*, 300). The strongest argument for the Copernican hypothesis was, of course, the simple explanation which it admitted of the retrogressions of the planets; further evidence is drawn from the newly discovered phases of the inferior planets, which show that these are dark bodies moving round the sun, and from the analogy with the heliocentric system presented by Jupiter and its train of satellites, the periods of which increase with increasing distance from the planet. The most serious objection to the Copernican theory was the non-appearance of annual stellar parallax. But Salviati proves that the stars may be more distant than was formerly supposed, their apparent size and nearness having been previously exaggerated. Simplicio asks what good purpose would be served by the interspace between Saturn and the stars. Salviati replies that the Creator may have had purposes in view other than man's welfare; or perhaps the interspace is occupied by undiscovered planets.

On the fourth and last day, Salviati explains his theory of the tides as depending upon the annual and diurnal motions of the earth. His theory is ingenious, but of no scientific significance, and we need not concern ourselves with it.

It is noteworthy that Galileo, in his Dialogue, ignores the cosmological system of Tycho Brahe as an alternative to the Ptolemaic theory alongside the Copernican. There are *two* chief systems, not *three*. Furthermore, he ignores the advance which Kepler had made on the circular planetary orbits of the ancients and of Copernicus, and he seems not to have mentioned the *Astronomia nova* anywhere in his writings; he says that his way of philosophizing is wholly different from Kepler's.

Galileo would have been justified in concluding his Dialogue with a decisive verdict in favor of the Copernican theory, based upon the weight of evidence in its support; but he was bound by the Edict of 1616 not to defend the theory. His book was accordingly written with obvious circumspection, and it was published in full compliance with the requirements of the ecclesiastical censorship. Immediately after the appearance of the book, however, proceedings were launched against Galileo on the ground that he had disobeyed a command, laid upon him in 1616, not to *teach* the Copernican theses condemned by the Inquisition in that year, and that he had deceived the Censor by concealing the fact of his having received such an injunction. Galileo, however, had no recollection of any such command. The minutes recording it seem to be an exaggeration of what took place when Galileo appeared before the papal authorities in 1616, or even, as E. Wohlwill was convinced, a forgery perpetrated in 1632 to provide evidence by which his condemnation could be ensured. Galileo's trial, in 1633, ended, as is well known, in his forced abjuration of the Copernican doctrines and in his spending the rest of his days as virtually a prisoner of the Inquisition at Arcetri, near Florence; he died in 1642. The ecclesiastical ban upon the *Dialogo,* the *De revolutionibus,* and three other Copernican works continued until 1822, when it was removed.

§ 7. RENÉ DESCARTES AND THE LAW OF INERTIA

Aristotle had conceived the celestial motions as representing a spontaneous activity of the planets or of their carrying-spheres.

About the beginning of the seventeenth century, however, the solar system began to be regarded as a mechanism the parts of which act upon one another according to universally valid physical laws; and so the fortunes of the Copernican theory came to be bound up with developments in the realm of dynamics.

Modern celestial mechanics may be said to have begun with the recognition that the motion of a cosmic body is conditioned by the agency of other bodies and not by the virtue of any ideal point in space. Even Copernicus found himself partly committed to this doctrine. Aristotle had taught that heavy bodies fell toward the center of the universe; the earth was formed incidentally by their concretion about that center. In the Copernican system, however, the motions of heavy bodies were seen to be directed toward the earth *as such,* and to vary from one instant to the next according to the position of the earth in its orbit. A new explanation of terrestrial gravity was therefore needed, which, as we have seen, Copernicus found in the affinity of parts to wholes. The moon could now no longer be regarded as circulating about the center of the universe; it typified a new class of bodies revolving about other bodies. For Copernicus the relation of a satellite to its primary was essentially a matter of geometrical order. It was Kepler who, in 1596, first roundly asserted the physical nullity of points in empty space (*Myst. cosmogr.,* cap. xvi; *Opera,* i, 158 ff.) and who began to assign to cosmic bodies an active role in determining the motions of which they are the goals or the centers. We have seen how he regarded the sun as the seat of an agency impelling the planets round their orbits. Terrestrial gravity he conceived as a "mutual corporeal affection between cognate bodies toward their union or connection" (*Astronomia nova,* 1609, Introductio; *Opera,* iii, 150 f), embracing the earth and the moon and giving rise to the tides, but not extending to the other cosmic bodies. Gravity for Kepler was an agency of the same order as magnetism; and magnetism figured also in his explanation of the sun's action upon a planet.

However, Kepler was precluded by the traditional doctrines of mechanics to which he adhered from classing together the earth's pull upon the moon and the sun's actions upon the planets; or, to put it more generally, he missed the essential connection between his two complementary conceptions of a celestial body as a goal and as a center of motion. He expected the primary body to administer to its satellite a continuous thrust in a direction generally tangential to the orbit of the latter. It had yet to be recognized that a planet, once set in motion, would move off of itself along a tangent unless it was retained in a closed orbit by a radial force directed toward the primary body. This revolution in outlook upon the planetary problem was effected during the middle years of the seventeenth century with the establishment of the law of inertia, which received its classic formulation in Newton's first law of motion.

Medieval cosmology had inherited from Aristotle the conception of space as an entity endowed with directive properties, determining the motions of the several elements each to its appointed place, or about the universal center, and so maintaining order in the world. Bruno's disintegration of the cosmos into an infinite space which could possess no privileged center, and Kepler's doctrine that only bodies could determine the motions of other bodies—such developments as these opened the way for reducing the role of space to that of a mere receptacle possessing only abstract geometrical properties. It was natural to ask what would be the motion, in such an abstract space, of a body isolated from the action of other bodies. What appears to be the earliest explicit and general formulation of the law of inertia occurs in the writings of the French philosopher René Descartes (1596–1650). He contributed more obviously to the general acceptance of the heliocentric theory by embodying it in a comprehensive system of natural philosophy which enjoyed a vogue lasting into the eighteenth century.

It was in 1633, soon after he had secluded himself in Holland for a life of meditation, that Descartes produced his first Copernican treatise; upon hearing of the condemnation of Galileo, however, he decided, as a devout Catholic, not to publish it, and it first appeared in print fourteen years after his death (*Le Monde de M. Descartes*, Paris, 1664; *Œuvres*, ed. Adam and Tannery, xi, 1).

Meanwhile Descartes had reformulated his system in greater detail and with the necessary accommodations to orthodoxy in his *Principes de la philosophie* (Paris, 1647; *Œuvres*, ix, 1). Under the guise of a fictitious world such as God *could* have created, Descartes conceived the primordial condition of things as a space of indefinite extent completely filled with a homogeneous matter, characterized only by spatial extension and the capacity for motion. At the Creation, God divided this continuum into particles of arbitrary size and shape and set these moving under laws of conservation which (among other provisions) ensure that a body, once projected, shall continue to move uniformly in a straight line unless forced to change its state by encounter with other bodies (*Principes*, II, 39). No void being admissible, all motion must consist of a circulation of matter; and the universe soon broke up into a plurality of systems of particles, each system revolving round its own center. At the centers of these *vortices* the finest and most rapidly moving particles (of which fire consists) collected to constitute stars, among which the sun is to be numbered; each of these exerts an outward pressure (perceived as light) which keeps its vortex in being against the opposing pressures of neighboring vortices. When a star becomes clogged with gross matter its light fades and its vortex collapses, and the defunct star is captured by a neighboring vortex, where it becomes a planet carried round the central star by the revolving stream of matter. The earth (with the other planets) originated in this manner; it is still the center of a miniature vortex, the mechanical cause of terrestrial gravity.

Thus the fictitious world of Descartes broadly embodied the Copernican scheme of the solar system. Motion, for Descartes, meant the displacement of a body relative to the matter in contact with it. But the earth, he maintained, did not move relative to the stream of particles surrounding it; therefore it was stationary, and no philosophical or Scriptural objections could arise. The Cartesian synthesis was of no lasting scientific value; its significance for us lies in its explicit formulation of the law of inertia and in its having served to introduce a wide circle of seventeenth-century readers to a form of the heliocentric theory.

§ 8. ISAAC NEWTON AND UNIVERSAL GRAVITATION

Descartes applied the law of inertia to the classic problem of the stone in the sling. The stone everywhere tends to move along the tangent to the circle it is describing; but this tangential motion is curbed by the tension in the cord, acting radially inward. In like manner, a revolving planet must be kept in its orbit by a force directed toward the center of motion. This principle came to be generally understood in the latter part of the seventeenth century; in particular it was grasped by the two English natural philosophers Isaac Newton (1642–1727) and Robert Hooke (1635–1703), whose closely related contributions to the planetary problem mark the culmination of the process we have endeavored to trace in this chapter. It was Newton who formulated the principles of classical mechanics and who applied them to the concrete problems presented by the solar system. He showed that Kepler's laws could be regarded as expressing certain properties of the orbits of planets moving under atractive forces directed toward the sun, and varying inversely as the square of the distance from that body. He further accounted for these forces by generalizing terrestrial gravity into a universal property of bodies. He thus established the heliocentric theory as the accepted cosmology of the modern world by showing how it opened the way to the interpretation of the laws of planetary motion as rational consequences of a few simple dynamical generalizations covering vast fields of observable fact.

By the middle of the seventeenth century, the heliocentric theory was widely accepted in Protestant England. In one of the notebooks which Newton began to keep as a boy there is an entry of two pages relating to Copernican astronomy; from the handwriting, David Eugene Smith assigned this portion to Newton's first year at Cambridge (*Isaac Newton, 1642–1727,* ed. W. J. Greenstreet, London, 1927, pp. 16 ff.). At Cambridge, too, Newton would have occasion to study the cosmology of Descartes, which his teacher, Isaac Barrow, had been largely instrumental in introducing there, and which was generally regarded as affording a rational basis for the heliocentric theory. However, Newton himself dated his speculations on the cosmic role of gravity and the mechanics of planetary motion from the "Plague Years" 1665–66, when the precautionary closing of Cambridge University drove him into seclusion at his Lincolnshire home. (For memoranda and correspondence illustrating the growth of Newton's gravitational theory see W. W. Rouse Ball, *An Essay on Newton's Principia,* London, 1893).

It appears that Newton began by wondering whether the influence of gravity might extend as far as the moon and, if so, whether it might not be the force which kept the moon in its orbit. To this context belongs the well-authenticated story that it was the fall of an apple in his orchard which brought dramatically to Newton's attention the age-old problem of gravity. He was aware that a body revolving at a uniform rate in a circle must be acted upon by a force accelerating the body toward the center; and he correctly formulated the relation connecting this acceleration with the radius of the circle described and the period of revolution. Hence, knowing the moon's approximate distance from the earth and the length of the sidereal month,

Newton was able to arrive at the moon's acceleration toward the earth (assuming the moon's orbit to be a circle uniformly described). On the other hand, supposing the planets likewise to describe circular orbits under forces directed toward the sun, Newton was able to deduce from Kepler's third law that these forces must be such as to determine accelerations inversely proportional to the squares of the distances of the several planets from the sun. He now assumed tentatively that the force of terrestrial gravity fell off with increase of distance from the center of the earth, according to this same law of the inverse square; thence, taking the moon's distance as sixty times the earth's radius, he inferred that the acceleration of a falling body at the earth's surface should be 3600 times the acceleration previously calculated for the moon. And when he compared this result with the acceleration actually measured on the earth, he "found them answer pretty nearly." More than twelve years were to elapse before Newton carried the problem any further. The explanation of this delay was probably a theoretical uncertainty as to whether, in comparing the distances, say, of the falling moon and of a falling apple from the earth, it was equally permissible in the two instances to reckon these distances from the earth's *center*, or whether this could be done only when the distance of the attracted body was large in comparison with the earth's radius. Newton did not clear up this problem until 1685. Meanwhile, in 1679, he had entered upon the next stage of his advance toward a mechanical theory of the solar system. This came about through a correspondence with Robert Hooke, who had himself already made significant and independent contributions to the planetary problem.

Hooke's tract *Cometa* (London, 1678), based in part upon notes of lectures which he delivered early in 1665, reflects his views on gravitation at a period antedating Newton's Lincolnshire speculations. Discussing the motions of comets, Hooke writes, "I suppose the gravitating power of the Sun in the center of this part of the Heaven in which we are, hath an attractive power upon all the bodies of the Planets and of the Earth that move about it, and that each of those again have a respect answerable, whereby they may be said to attract the Sun in the same manner as the Load-stone [magnet] hath to Iron, and the Iron hath to the Load-stone" (*Cometa*, 12). Hooke's comparison of gravitation to magnetism links his ideas with those of Gilbert and Kepler. It naturally suggested to him that the force of terrestrial gravity should fall off with increase of distance from the earth, a phenomenon which he sought in vain to establish by ingenious experiments from 1662. In his *Micrographia*, published in January 1665 and widely circulated, Hooke referred again to this "decrease of the power of gravity" with increasing distance from the earth, although without suggesting any precise law of variation. In 1666 Hooke explained to the Royal Society his hypothesis of an attractive force exerted by the sun upon all surrounding bodies, constraining the planets to describe closed curves instead of traveling off along tangents into space. He staged a rough and ready demonstration of a planet's motion with the aid of a "conical pendulum"—a weight suspended by a cord from a fixed support and revolving in a circle under the inward force provided by the tension in the cord. In 1669 he set himself to observe a star systematically for annual parallax, employing a species of telescope equipped with a micrometer eyepiece. He claimed (mistakenly) to have established this "confirmation of the Copernican System"; and in the closing passages of his descriptive lecture (1670) he returned to the theme of a system of mutually attractive heavenly bodies.

It was toward the close of 1679 that Hooke entered into a correspondence with Newton in the course of which the latter predicted that a diurnal (eastward) rotation of the earth should make a freely falling body show an eastward deviation from the vertical—a phenomenon which Hooke promptly (and overoptimistically) claimed to have established by experiment. Hooke and Newton continued for some little time to discuss what path a heavy body would pursue if it could fall toward the earth's center without encountering any obstruction. Hooke postulated that, at points above ground level, the force of gravity should vary inversely as the square of the distance from the earth's center. He urged Newton to investigate mathematically what would be the curve described by a particle under a central force subject to this law. It must have been in the period immediately following that Newton succeeded in connecting the elliptic orbit of a planet with the inverse-square law of gravitation. He kept his discovery to himself and even lost his calculations; it was not until August 1684 that Halley recalled his attention to the problem and induced him to embark upon the researches leading to the publication of the *Principia,* in 1687.

In that historic work, Newton finally broke with the ancient doctrines of privileged points, natural places, affinities, and so forth, assigning instead to the mutual attractions of cosmic bodies the role of altering their states of rest or of uniform motion. The laws of Kepler received general formulation in mechanical terms in Book I; this book contains also an approximate analysis of the problem of three mutually gravitating bodies for a particular case illustrating the behavior of the moon under the combined attractions of the earth and the sun. Here, too, the slow conical motion of the earth's axis, which Copernicus had postulated to account for precession (see Chapter IV, § 2, *supra*), receives for the first time a mechanical explanation as arising from asymmetrical forces exerted by the sun and the moon upon the earth's equatorial protuberance. A section is devoted to establishing the fundamental proposition that two homogeneous spheres attract each other like massive particles concentrated at their respective centers.

Of the propositions in Book II relating to motion in resisting media, cosmological significance chiefly attaches to those which establish that the Cartesian hypothesis of vortices cannot be reconciled with Kepler's laws nor with certain other results of observation.

In the third and last book of his *Principia*, Newton makes concrete application of the foregoing results to "demonstrate the frame of the System of the World" under the generalization of universal gravitation: "That there is a power of gravity pertaining to all bodies, proportional to the several quantities of matter which they contain," and that "the force of gravity towards the several equal particles of any body is inversely as the square of the distance of places from the particles" (Book III, Prop. VII). In consequence of the equality of action and reaction postulated in Newton's third law of motion, the sun is attracted by each planet and "must continually be moved every way" (Prop. XII). Hence the center of the universe, which is, by hypothesis, a stationary and immovable point, could not be assumed to coincide with the center of the sun, as in a literally heliocentric system. Instead, Newton identified the center of the universe with the "center of gravity" (mass center) of the solar system, from which, however, the sun never departs by more than about the length of its own diameter. Newton assumed the mass center of the solar system to be immovable; and it was left for Herschel to prove, a century later, that the sun and planets as a whole were traveling through space in relation to the surrounding stars.

With Newton we reach the natural conclusion of these studies in the history of the Copernican doctrine. The heliocentric theory was finally established through the general acceptance of the system of which that theory formed an integral part. Voices were, indeed, raised occasionally in behalf of the geocentric theory during the eighteenth century, and even later; but such opinions are now to be found only in fanatical circles. It remained for astronomers of the generations following Newton to prove that his mechanical principles sufficed to account for the intricacies in the motions of the celestial bodies down to all but the last detail. Only an unexplained discrepancy in the rate of progression of the perihelion of Mercury remained outstanding to be the Achilles' heel of the Newtonian theory of gravitation, and to constitute, in due course, the crucial test of a still more comprehensive synthesis.

8

The Physical Verification of the Copernican Theory

THE DIURNAL AND ANNUAL MOTIONS WHICH COPERNICUS ASSIGNED
to the earth were not directly perceptible; they could be verified only indirectly by a comparison of their necessary consequences with the results of observation or of deliberately devised experiment. To Copernicus himself and his earliest disciples the heliocentric theory was commended by the economy
with which it accounted for the major celestial phenomena; the
natural philosophers of his day could contend only that it was
fundamentally incompatible with accepted physical doctrines.
In the seventeenth century, however, the reformation of physics
and the invention of the telescope appeared to open up possi-
bilities of confirming the motion of the earth by the detection
of certain minute effects which might be expected to result
therefrom. In these closing pages we shall glance at some of the
investigations undertaken, and the considerations advanced, to
"prove" the Copernican theory, tracing the developments and
diversifications of method to the early years of the present century, when the age-long controversy concerning the rest or motion of cosmic bodies came to be viewed in an entirely new
light. We shall consider the quest for evidence first of the *diurnal rotation* of the earth (§§ 1–4) and then of its *annual revolution* (§§ 5, 6).

§ 1. THE DEVIATION OF FALLING BODIES

Several of the classic arguments on both sides of the Copernican controversy were based upon assumptions as to the effect
which a rotation of the earth would produce upon the apparent
motion of a projectile or freely falling body. We have already
encountered Tycho Brahe's argument that an eastward rotation of the earth would give a cannon a greater range toward
the west than toward the east. In the sixteenth century it was
still generally assumed that such a rotation would produce also
an apparent westward deflection in a falling body. Instead of
striking the ground vertically below its point of release, the
body would appear to fall to the westward by an amount equal
to the eastward travel of the earth's surface during the time of
fall. The observed absence of any such effect in a body thrown
vertically upward was advanced by Aristotle and his successors
as proof that the earth does not rotate axially.

Copernicus and his early disciples met this argument by supposing the falling body to partake of the same natural circular
motion as the rest of the earth, and thus to keep up with the
latter in its eastward rotation (see Chapter III, § 3, *supra*).
Thomas Digges and Giordano Bruno, whose contributions to
the spread of the Copernican doctrine we have already considered, each sought to illustrate this principle by reference to the
ideal experiment of lowering or dropping a weight from the
masthead of a ship; however fast the ship is sailing, the body
(they contended) must appear, to anyone on board, to descend
perpendicularly to the foot of the mast. We encounter the same
analogy of a falling body upon a moving ship in Tycho Brahe's
notes on his correspondence (already mentioned) with Christoph Rothmann. Here, however, Tycho is arguing against the
diurnal rotation; and he contrasts the fall of a bullet from a lofty
tower to a point vertically below with the behavior of a missile
thrown upward from the deck of a ship, which (he maintains)

will fall toward the stern, and the more so the greater the speed of the ship. The same problem is discussed again, but with deeper insight into mechanical principles, by Galileo in his *Dialogo* of 1632, where Simplicio adopts Tycho Brahe's line of argument but Salviati retorts that "one cannot from the Stone's falling perpendicularly at the foot of the Tower, conclude anything touching the motion or rest of the Earth" (Salusbury's translation, 1661, 126). Salviati's assertion was put to the test of actual experiment in various ways, notably by Marin Mersenne and Pierre Petit, who, in 1634, at the instigation of Descartes, fired balls vertically upward to see whether they would fall back into the cannon's mouth. The results were disappointing; but by the middle of the seventeenth century it was generally agreed among the more progressive natural philosophers that no observations of the behavior of falling bodies could decide the question of the earth's rest or motion. To quote Galileo again: "All Experiments that can be made upon the Earth are insufficient means to conclude its Mobility, but are indifferently applicable to the Earth movable or immovable" (Introduction to *Two Chief Systems*, trans. T. Salusbury).

The dramatic new turn given to the problem by Newton in predicting an *eastward* fall of bodies has already been noted at the point where it became relevant to our account of the development of planetary theory (see Chapter VII, § 8, *supra*). If we imagine a heavy body let fall, "its gravity (wrote Newton) will give it a new motion towards the center of the Earth without diminishing the old one from west to east. Whence the motion of this body from west to east, by reason that before it fell it was more distant from the center of the Earth than the parts of the Earth at which it arrives in its fall, will be greater than the motion from west to east of [those] parts of the Earth . . . and therefore it will not descend the perpendicular . . . but outrunning the parts of the Earth will shoot forward to the east side of the perpendicular." (W. W. Rouse Ball, *An Essay on Newton's Principia*, 1893, p. 143.)

In the experiments by which he claimed to have demonstrated this excessively minute effect, Robert Hooke neglected the precautions found indispensable by later investigators who have progressively refined upon his crude technique. Following Hooke, the pioneer workers were G. B. Guglielmini, who experimented in 1791 by dropping leaden balls in the Asinella Tower at Bologna, and J. F. Benzenberg, who utilized for the purpose the tower of St. Michael's Church in Hamburg (1802) and a mine-shaft in Schlebusch (1804). Some idea of the order of magnitude of the effect is afforded by Benzenberg's results in Hamburg, where 31 balls, dropped from a height of 235 feet, showed a mean eastward deviation of only one-third of an inch and a mean southward deviation of one-eighth of an inch. The phenomenon was subsequently investigated by F. Reich (1831), of Freiberg, and E. H. Hall (1902), of Harvard. Attempts to establish a theoretical formula for the deviations were begun in the early years of the nineteenth century by Gauss and Laplace. The results of the experiments, on the whole, confirmed Newton's surmise by establishing an eastward displacement of falling bodies of the theoretically indicated order of magnitude; but there were also persistent suggestions of a much smaller but still unaccountably large southward component—which indeed, Hooke also claimed to have detected. In 1912, J. G. Hagen, of the Vatican Observatory, carried out experiments on the eastward deviation of the descending weight of an Atwood's machine. He obtained a positive result of the expected order; but no measurable southward deviation was established. (See J. G. Hagen, "La Rotation de la Terre; ses Preuves mécaniques anciennes et nouvelles," *Pubbl. d. Specola Astronomica Vaticana,* Ser. 2, 1912, i, where other devices and proposals for demonstrating the earth's rotation mechanically are described; and A.

Armitage, "The Deviation of Falling Bodies," *Annals of Science*, 1947, v, 342.)

§ 2. THE VARIATION OF GRAVITY AND THE FIGURE OF THE EARTH

Another physical argument for the diurnal rotation was afforded by the discovery that the force of gravity varies slightly over the surface of the earth, being less in lower than in higher latitudes. The great seventeenth-century Dutch physicist Christian Huygens, a Copernican, applied his theory of the so-called "centrifugal force" to calculate the diminution of gravity and the deflection of the plumb line which should be produced in any given latitude by such a diurnal rotation; and he drew verifiable conclusions as to the probable shape, or figure, of the earth. As early as 1659 he calculated that the centrifugal force upon a body at the equator should amount to $\frac{1}{265}$ of its weight (*Œuvres complètes*, xvi, 304; Newton gave the more accurate fraction $\frac{1}{289}$). Huygens' theory of the pendulum and his invention of the pendulum clock provided the means of comparing gravity in different latitudes with great precision. In 1666 he calculated how much slower such a clock should go on the equator than at the poles or in latitude 45° (*ibid.*, xvii, 285 f.); and he recommended observing the rates of a clock in various latitudes as a means of verifying the earth's rotation. Then, in 1672, Jean Richer found the length of the seconds pendulum shorter at Cayenne (in latitude 5° N. approximately) than at Paris; and similar comparisons subsequently made in various parts of the world conformed to a general rule. Huygens naturally linked this phenomenon with the earth's rotation; and by the beginning of 1687 (before seeing Newton's *Principia*) he had drawn certain inferences as to the shape of the earth which he appended to his *Discours de la cause de la pesanteur* (originally delivered in 1669 and eventually published in 1690; *ibid.*, xxi, 443).

Huygens considered the bob of a plumb line as being kept in equilibrium under three forces: (a) the force of gravity, acting through the earth's center; (b) the centrifugal force, or the tendency of the bob to fly off radially from the earth's axis of rotation; and (c) the tension in the plumb line. He showed that there must, in general, be a deflection of the plumb line, so that this would not point to the center of the earth. Solving a triangle of forces, he found that this deflection would amount, in the latitude of Paris, to nearly six minutes of arc. Since the surface of a liquid at rest is everywhere perpendicular to the plumb line, it followed from the deflection of the latter that the surface of the ocean could not be that of a sphere but rather that of a spheroid flattened at the poles. And since the general level of the land does not differ markedly from that of the ocean, the whole surface of the globe must approximate to the same figure, which Huygens supposed must be roughly such as an ellipse would generate by turning about its minor axis.

In 1687 Huygens received a copy of Newton's newly published *Principia* as a gift from its author; and his ideas underwent a further development. Newton argued (*Principia*, Book III, Prop. XVIII) that, if the earth were supposed to be in rotation, it could not be an exact sphere, or the ocean, in its tendency to fly outward from the axis of rotation, would accumulate at the equator and submerge the land there. He inferred that the earth, under this same centrifugal tendency, must have assumed a spheroidal form, with its polar axis less than its equatorial axis; and he sought (Prop. XIX) to calculate the ratio of these axes. As early as 1683 Robert Hooke had drawn conclusions as to the figure of the earth from the supposed equilibrium of two ideal columns or cones of liquid lying respectively along the polar axis and in the equatorial plane and communicating at the center of the earth. Newton treated the earth as a homogeneous mass of mutually gravitating fluid. He, too, considered two communicating columns, or "canals," of this fluid, one extending from one pole to the center of the earth and the other from the center to a point upon the equator; he assumed that these columns must just balance each other, account being taken of the attraction of each toward the center, and of the centrifugal tendency of the equatorial column, which must just compensate for its greater length and weight. Newton thus arrived at an estimate of the ellipticity of a meridian section of the earth. He proceeded (Prop. XX) to formulate the law according to which gravity should increase with latitude on the spheroidal earth; his formula (which was confirmed in principle by the more refined theoretical investigation of Clairaut, in the eighteenth century) showed a fair measure of agreement with the available observations made in various parts of the world.

After reading Newton's *Principia*, Huygens made a further addition to his *Discours*. He applied Newton's artifice of communicating canals in an attempt to determine the exact figure of the earth, conceived as entirely covered by a liquid surface everywhere perpendicular to the plumb line. Huygens denied universal gravitation, and his calculation was inferior to Newton's, but it gave the same general impression of the earth as a spheroid having its greatest diameter in the plane of the equator. These conclusions as to the shape of the earth were confirmed during the eighteenth century by the comparison of the measured lengths of a degree of meridian taken in high and low latitudes respectively.

The establishment of the spheroidal figure of the earth, and of the variation of gravity over its surface, was regarded as affording indirect verification of the earth's diurnal rotation, from which, on Newtonian principles, these phenomena could be deduced as necessary consequences. To Hooke, Huygens, and Newton alike it appeared significant that, in the planet Jupiter, a rapid axial rotation was known to be associated with an appreciable equatorial protuberance. It may be noted also that Newton's conclusion as to the shape of the earth, combined with the hypothesis of its rotation, enabled him to give a dynamical explanation of the precession of the equinoxes (Chapter IV, § 2, *supra*) as arising from the couples exerted upon the earth's protuberance by the sun and moon.

§ 3. THE CIRCULATION OF THE ATMOSPHERE AND FOUCAULT'S EXPERIMENTS

During the seventeenth century attempts were made to account for the northeast and southeast trade winds of the tropical ocean by reference to the diurnal rotation of the earth. Galileo suggested that the general westward flow of air in the tropics represented a lagging of the atmosphere behind the rapid eastward motion of the earth's surface (*Dialogo,* fourth day). Later it was recognized that an important part was played by the heat of the sun, which must dilate and rarefy the air round the equatorial belt and cause it to ascend, giving place to an inflow of colder air which constituted the trade winds. The fact that these winds did not blow due north and south, but showed a westward deflection, was at first attributed to their supposed tendency to follow the subsolar point as it moves daily round the earth. The true bearing of the earth's rotation upon the apparent directions of the permanent winds was explained, in 1735, by George Hadley (*Phil. Trans.,* 1735, xxxix, 58). He pointed out that, since the eastward speed of the earth's surface increases progressively from the pole to the equator, the wind blowing southward to take the place of the rarefied air "having a less Velocity than the Parts of the Earth it arrives at, will have a relative Motion contrary to that of the diurnal Motion of the Earth in those Parts, which being combined with the Motion towards the Equator, a north-east Wind will be produced on this Side of the Equator, and a south-east on the other," while in our temperate zone the winds blowing northward must appear to be progressively deflected eastward. Later (1856) William

Ferrel formulated more generally the principle by which the characteristic bias of the trade winds, the southwesterly winds of the northern hemisphere, and the "Brave West Winds" of the southern, are all attributed to the diurnal rotation of the earth, as are also the rules governing the directions of air flow in cyclones and anti-cyclones.

The diurnal rotation should also (according to Poisson, 1837) produce an analogous effect upon the apparent motion of a projectile fired horizontally; in the northern hemisphere it should appear to deviate toward the right of the target at which it is aimed, and in the southern hemisphere, toward the left. This effect, for which allowance has to be made in gunnery, is barely perceptible. However, a century ago the French physicist Léon Foucault devised a simple experiment which he conceived as a means of *accumulating* the deviations of a projectile discharged backward and forward along its trajectory, and thus of demonstrating the diurnal rotation of the earth. He set a simple pendulum swinging in a well-defined plane of oscillation, and he established that this plane rotated about the vertical, relative to the surrounding objects. Its rate of rotation proved to be that of the hypothetical rotation of the earth multiplied by the sine of the latitude of the place of observation; the earth's diurnal motion could be plausibly inferred from the phenomenon. Foucault's first successful trials (carried out in a cellar under his house in Paris) were announced on February 3, 1851; his most spectacular demonstration was staged in the Panthéon. The experiment was repeated in many quarters, and it was interpreted on classical mechanical principles.

Foucault's success was followed up by Bravais, who used a conical pendulum (in which the bob described a circle at a rate hastened or retarded by the earth's rotation), by Hengler, who used a horizontal pendulum, and by other investigators with various forms of compound pendulum. The intricacies of pendular motion upon the rotating earth (unsuspected by Foucault) were explored theoretically and experimentally by H. Kamerlingh Onnes in a classic thesis presented at Groningen in 1879 (and summarized by J. Stein in *Pubbl. d. Specola Astronomica Vaticana,* Ser. 2, 1912, i, Appendice).

Meanwhile, Foucault had recognized another means of demonstrating the rotation of the earth in the mechanical properties possessed by a body rotating about an axis of symmetry; he employed a wheel with a massive bronze rim and a steel axle, mounted so as to turn freely about its center of gravity. Such an apparatus, employed for this purpose, he termed a *gyroscope.* Its significant properties are that the axis of rotation, when unconstrained, maintains a fixed direction in space and that it responds in a characteristic manner to forces tending to change that direction. The possibility of employing the fixity of direction of the axis of a rotating body to demonstrate the earth's diurnal motion had been discussed by others before Foucault, but it was first realized successfully by him.

Foucault performed three experiments the results of which could all be explained dynamically by reference to a rotation of the earth about its polar axis: (*a*) If the gyroscope was set in rotation with its axis free to turn in both a horizontal and a vertical plane, it was observed to follow the celestial sphere in its diurnal motion (i.e., its axis continued to point to the same star, after the manner of a clock-driven equatorial telescope). (*b*) If the axis of the gyroscope was free to turn only in the horizontal plane, it moved round so as to set itself due north and south. (*c*) Finally, if the axis of rotation was free to turn only in the plane of the meridian, it tended to set itself parallel to the polar axis in such a manner that the sense of its rotation should be the same as that ascribed to the earth. The gyro-compass, employed in navigation by sea and air in place of the magnetic compass, utilizes the fundamental properties of the gyroscope; it is controlled by the rotation of the earth. (See *Recueil des travaux scientifiques de Léon Foucault,* Paris, 1878.)

§ 4. THE VARIATION OF LATITUDE

In the eighteenth century Leonhard Euler inferred from dynamical considerations that a diurnal rotation of the earth should be associated with a periodic variation in the latitudes of all places upon its surface (*Mém. de l'Acad. Roy. des Sciences,* Berlin, *année* 1758, xiv, 165, 194 ff.). If the earth is assumed to be a spheroid slightly flattened at the poles and with its axes of rotation and of symmetry not quite coincident, the axis of symmetry should describe a narrow cone about the axis of rotation in a period which Euler put at about 305 days. In consequence, the north and south terrestrial poles (in which the axis of rotation cuts the earth's surface) must ideally describe small circles about their mean positions, thereby producing in the latitudes of places corresponding fluctuations the detection of which could be regarded as an indirect verification of the earth's rotation. Evidence of minute variations in the latitudes of several observatories was obtained about the middle of the nineteenth century; the effect was established independently by S. C. Chandler, of Cambridge, Mass., and F. Küstner, of Berlin. Chandler's analysis of the results (1891) revealed that each terrestrial pole revolved about a mean position, with a radius of the order of 30 feet, in a period of about 427 days, the discrepancy between the predicted and the actual periods being attributed by Simon Newcomb to the nonrigidity of the earth. Over and above this "Eulerian nutation" there are further variations of latitude of period one year, supposedly due to meteorological causes.

§ 5. THE ABERRATION OF LIGHT

Repeated reference has been made above to the annual stellar parallax which should result from an orbital revolution of the earth about the sun (see Chapter III, § 4; Chapter VI, § 2, Chapter VII, §§ 4, 6, 8). Prominent among those who searched for stellar parallax in the eighteenth century was James Bradley (*c*. 1693–1762), an observer of genius who became Astronomer Royal in 1742. Bradley followed Hooke's procedure of endeavoring to detect seasonal fluctuations, due to parallax, in the meridian altitude of the star γ Draconis (*Phil. Trans.*, 1728, pp. 637 ff.), employing for this purpose an instrument of a type known as a zenith sector. It was in December 1725 that Bradley began to make systematic measurements of the altitude at which the selected star crossed the meridian. It was to be expected that annual parallax would make the meridian altitude of γ Draconis a maximum in June and a minimum in December. An annual fluctuation was, in fact, detected; but the maximum and minimum altitudes fell in September and March respectively, so that the effect could not be attributed to parallax. The altitudes of a number of other stars transiting near the zenith were found to show analogous fluctuations, which, however, differed from one another in extent and in phase according to the longitudes and latitudes of the individual stars. Several explanatory hypotheses were suggested and disproved; but at length, in 1728, Bradley hit upon a satisfactory explanation of the phenomenon and its laws, involving the hypothesis of the earth's annual revolution and based upon the assumption that the ratio of the earth's orbital velocity to that of the propagation of light is not negligible. In consequence of this *aberration* of light (as the phenomenon came to be called), each star must annually appear to describe, about its mean position upon the celestial sphere, a small ellipse the major axis of which lies parallel to the ecliptic and subtends about 40″, while its eccentricity depends upon the latitude of the star. The results of Bradley's observations conformed to this deduction; they therefore constituted an argument for the truth of the Copernican theory as impressive as would have been the discovery of the stellar parallax which his prolonged search failed to reveal. For aberration is hardly less immediately deducible than parallax from the hypothesis of the earth's orbital motion.

§ 6. ANNUAL STELLAR PARALLAX

Bradley was not the only observer who followed the pioneer venture of Robert Hooke in applying his telescope to the quest for the elusive phenomenon of stellar parallax; but no substantial results were obtained until 1838, when the German astronomer F. W. Bessel (1784–1846) detected the parallax of a star and measured it with reasonable accuracy (*Astronomische Nachrichten*, Dec. 13, 1838, Nos. 365, 366; 1840, Nos. 401, 402). The parallactic displacement may be expected to be most conspicuous in those stars which lie nearest to the solar system. It was at first mistakenly assumed that the nearest stars must be the brightest; but Bessel adopted as the criterion of proximity the rapidity of a star's proper motion on the celestial sphere; and he was led to select for examination a double star of the fifth magnitude known as 61 Cygni. His procedure was to measure, by means of a micrometer of the utmost refinement, the angular distance of this object (or, more precisely, of the point midway between its components) from two neighboring faint stars the minuteness of the proper motions of which afforded grounds for supposing that their parallaxes would be insensible, so that the *differential* parallax, measured between the selected star and these neighbors, could reasonably be identified with the required *absolute* parallax. (This procedure had been clearly outlined by Galileo in his *Dialogo* of 1632.) Bessel's main series of observations extended over a whole year (beginning in August 1837), and they were further continued in 1839–40. When reduced with all necessary allowances for astronomical and instrumental factors, the observations indicated an annual parallax of about 0".35 for 61 Cygni, corresponding to a distance of the star from us of the order of 60 million million miles.

Bessel was anticipated in the discovery of stellar parallax by a Scotsman, Thomas Henderson, whose results, however, were first made public some weeks after those of the German astronomer. Henderson's observations related to the star α Centauri, one of the nearest known stellar neighbors of the sun.

The phenomenon of annual stellar parallax, established (latterly with the aid of photography) for hundreds of stars, constitutes a classic proof of the earth's orbital motion; the only alternative hypothesis would involve the revolution of the stars themselves in annual orbits in phase with the sun.

Epilogue

WITH THE FOUNDATION OF MODERN ASTRONOMY UPON THE HELIO-
centric theory, we reach the limits of the field which it was proposed to cover in this book. However, we cannot ignore completely the subsequent developments in scientific thought which
have altered the status of the historic planetary theories and set
the Copernican controversy in a new light.

In the middle of the sixteenth century the universe was still
conceived as a finite space; its boundary constituted a frame of
reference defining the absolute rest or motion of objects; and an
obvious meaning could be attached to the assertion that some
specified body, whether the earth or the sun, occupied the privileged central position. With the breakup of the medieval cosmology, space came to be conceived as infinite, or at least as
indefinitely extended. Where there was no center, there could
be no central body; and a new criterion of absolute rest or motion had to be sought. Newton found it difficult to conceive how
absolute linear motion could be measured; but he postulated
the existence of an absolute space relative to which the "center
of gravity" (mass center) of the solar system is at rest. However,
with the eighteenth century came Edmond Halley's announce-
ment (1718) of his discovery that certain stars are apparently
moving freely through space in various directions, and Sir William Herschel's proof (1783) that the sun, with its train of planets, is traveling through the heavens like any other star. It still
seemed reasonable to assume that the intrinsic motions of the
stars would be random, so that a standard frame of reference
might be defined by the centroid of the stellar system. This view
was rendered untenable, however, by the discovery of the phe-

nomenon of star-streaming, announced by J. C. Kapteyn in 1904. Meanwhile, in a renewed search for distinguishable "landmarks" in space, attempts had been made to measure, by means of appropriate optical methods, the earth's velocity relative to an ether conceived as a stationary medium filling the whole of space; but the negative results of such experiments as those carried out by A. A. Michelson and E. W. Morley in 1887 dispelled hopes of progress along these lines.

Newton, although he admitted the difficulties in the way of determining an absolute *linear* velocity, had deemed it feasible to ascertain the absolute *angular* velocity of a rotating body by measuring the centrifugal forces thereby called into play. Foucault's experiments also appeared to afford a means of demonstrating the absolute rotation of the earth. Ernst Mach maintained, however, that such inertial phenomena cannot constitute a criterion for deciding whether the diurnal motion belongs to the earth or to the rest of the universe.

By the end of the nineteenth century the problem had thus reached an *impasse,* of which, however, a radically new interpretation was soon afforded by Einstein's principle of relativity, with its denial that statements as to the absolute rest or motion of bodies correspond to any real distinction in nature. Henceforward, the decision to regard one cosmic body as fixed and another as moving could be related only to considerations of convenience in systematizing physical phenomena. It was the historic achievement of Copernicus that he chose to systematize planetary astronomy upon a principle which has opened for his successors a way to the unitary cosmology of today, with its wide and luminous horizon.

Notes

NOTE I. EQUIVALENCE OF ECCENTRIC AND EPICYCLIC SYSTEMS

To illustrate simply the equivalence of eccentric and epicyclic planetary systems, consider a point P uniformly describing a fixed eccentric circle APA′ with center C, the earth being at E (Fig. 40). Draw a circle, with center E, having the same radius

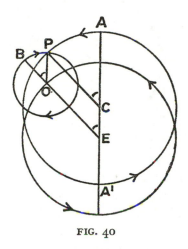

FIG. 40

as the eccentric, and draw the radius EO parallel to CP. Join OP. With center O and radius OP, draw a circle, and produce EO to meet it again at B. As P describes the eccentric, let EO swing round so as always to remain equal and parallel to CP, so that OP remains equal and parallel to EC. Then since ∠ACP = ∠AEO = ∠POB, OP swings round at a uniform rate relative to OB, and P describes an epicycle about O, while O uniformly describes a deferent about E. The same motion of P can thus be represented equally well on either system.

Frequent allusion is made in the foregoing pages to Copernicus' Table of Chords (*Canon subtensarum in circulo rectarum linearum*); this calls for a word of explanation. In Ptolemy's *Almagest* (I, 9) there is a table giving the lengths of the chord AB of a circle (Fig. 41) corresponding to certain assigned values

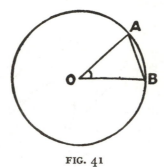

FIG. 41

of the angle AOB which that chord subtends at the center O. These lengths are expressed in terms of a unit, or "part," which is one-sixtieth of the radius; they are therefore independent of the dimensions of any particular circle. The chords of certain angles are known from elementary geometry (e.g., chord of 90° = 60 $\sqrt{2}$ parts). The calculation of the remainder of the table depends upon a set of theorems analogous to the formulae of elementary trigonometry, and of which Ptolemy gives proofs; he seems, however, to have derived both the table and the underlying propositions from Hipparchus of Rhodes. Such a "Table of Chords" was the Greek equivalent of a sine table, since

$$\text{chord of } \angle AOB = 2 \cdot AO.\sin \tfrac{1}{2} (\angle AOB)$$

The conception of the sine of an angle measured by half the chord of double the angle, appears to have been introduced by the Hindus not later than the fifth century A.D.; it was adopted by the Arabs, and a table of sines was included in the *Epitome in almagestum* of Purbach and Regiomontanus, a copy of which Copernicus acquired soon after its publication (1496).

Copernicus' table (I, 12) gives the semichord of twice the angle, for angles increasing by successive increments of 10′, from 0° to 90°, with a column of mean differences to facilitate interpolation. As Copernicus takes the hundred-thousandth part of the radius of the circle as his unit, or "part," the figures in his table are the same as those in a modern table of sines.

NOTE III. THE *Commentariolus* AND THE *Narratio prima*

The historic book in which Copernicus elaborated the details of his planetary system was not published until 1543, the year of its author's death. In the meantime, however, the essential doctrines of the book had twice been put into literary form to satisfy the curiosity which its fame had awakened. Copernicus himself drew up a brief, nontechnical account of his system, generally known as the *Commentariolus,* which appears to have been circulated in manuscript in sympathetic quarters. Later the main facts were made known to a wider public through the more detailed *Narratio prima,* which Rheticus published in 1540. We shall now indicate briefly the contents of these two compositions.

Two manuscripts are known to exist of the synopsis entitled *Nicolai Coppernici de hypothesibus motuum coelestium a se constitutis commentariolus.* The first of these was discovered at Vienna by M. Curtze and edited by him in 1878 (*Mittheilungen des Coppernicus-Vereins zu Thorn,* Heft I, 1–17). The second was found shortly afterward at Stockholm by A. Lindhagen (*Bihang till K. Svenska Vet. Akad. Handlingar,* Bd. 6, 1881).

The text was established from these two manuscripts by L. Prowe (*Nicolaus Coppernicus*, II, 184–202). It has been translated into English by E. Rosen in *Osiris*, III, Pt. 1, 1937, and in his *Three Copernican Treatises*, New York, 1939. Curtze assigned the composition of the *Commentariolus* to the period 1533–39. Birkenmajer, however, adduces evidence for supposing that its contents were elaborated before 1509 and that it was composed not later than 1512 (*Mikolaj Kopernik*, 1900). His reasons for assigning so early a date to the *Commentariolus* depend upon his contention (see Chapter II, § 4, *supra*) that Copernicus worked out two successive types of planetary theory, and that the one given in this little tract is the earlier of the two.

The *Commentariolus* begins with a brief allusion to the two typical planetary theories of antiquity and indicates their characteristic deficiencies. The system of homocentric spheres failed to account for the undoubted variations in the distances of the planets from the earth; the rival system of eccentrics and epicycles, as developed by Ptolemy, was free from this objection, but it involved assumptions which conflicted with sound physics. Copernicus, seeking a way out of this dilemma, at length found that a reasonable and satisfactory explanation of the planetary motions could be given, provided the following postulates were granted:

(*a*) That the celestial spheres have not all the same center.

(*b*) That the center of the earth is not the center of the universe but is only the center of gravity and of the moon's sphere.

(*c*) That the sun is at the center of all the planetary spheres and of the universe.

(*d*) That the ratio of the sun's distance from the earth to the height of the firmament is less than that of the earth's radius to the sun's distance and is therefore negligible.

(*e*) That all apparent motion in the firmament is due to the motion of the earth, which, with the surrounding elements, daily turns upon poles fixed in the firmament.

(*f*) That all apparent motion of the sun is due to the movement of the earth, which revolves about the sun like any other planet.

(*g*) That the progressions and retrogressions of the planets are to be attributed to the motion of the earth, which alone suffices to explain all the apparent non-uniformities in the heavens.

The technical details of the Copernican system are then summarized under seven heads, mathematical demonstrations "destined for a larger work" being omitted:

(*a*) The order of succession of the planetary spheres, with their approximate periods of revolution, is laid down in accordance with the heliocentric scheme of Fig. 7.

(*b*) The triple motion of the earth is indicated, and precession is attributed to a mutability in the direction of the earth's axis of rotation.

(*c*) It is recommended that the length of the year be deduced from observations of the apparent motion of the sun relative to the fixed stars and not to the equinoxes.

(*d*) A theory of the motion of the moon is outlined which is precisely that developed in Book IV of the *De revolutionibus* (see Chapter V, § 1, *supra*).

(*e*) A theory of the motions in longitude of Saturn, Jupiter, and Mars is here outlined; but it differs from that eventually adopted in Book V of the *De revolutionibus*. There each of these planets is supposed to describe an epicycle the center of which describes a deferent *eccentric* to the earth's orbit (see Chapter VI, § 2, *supra*). Here, in the *Commentariolus*, each planet's motion is represented by means of two epicycles, one superimposed upon the other, and a deferent *concentric* with the earth's orbit, the whole forming a system closely similar to that assigned to the moon. (This represents the earlier type of planetary theory, which Birkenmajer traced in canceled passages of the manuscript of the *De revolutionibus*.)

(f) The system of Venus is also constituted by a deferent and two epicycles, after the manner of the superior planets.

(g) The system of Mercury, conceived on similar lines, is, however, complicated by a periodic fluctuation in the radius of the deferent. The theories of the motions in latitude of the several planets, outlined under (e), (f), and (g), resemble those developed in Book VI of the De revolutionibus (see Chapter VI, § 4, supra).

Copernicus finally reckons up the number of separate circular motions involved, explicitly or implicitly, in the foregoing hypotheses; they amount to 34.

The Narratio prima is concerned mainly with the contents of Book III of the De revolutionibus; Rheticus had not yet completed his study of the manuscript of Copernicus when he addressed this report to his old teacher, Johann Schöner, and he intended to follow it up with further Narrationes. Besides being a competent astronomer, Rheticus was an accomplished classical scholar, and he writes in a somewhat flowery style, with many high-flown, although doubtless sincerely meant, panegyrics on his "Dominus Doctor," Copernicus, whom, however, he does not mention by name. He treats the various topics in a rather different order from that in which they occur in the manuscript. After briefly indicating the contents of all six books, he deals, in succession, with the determination of the rate of precession and of the amplitude and period of its anomaly; with the measurement of the tropical and sidereal years; and with the periodic changes in the obliquity of the ecliptic and in the eccentricity and apse line of the earth's orbit (see Chapter IV, supra).

At this point Rheticus turns aside for a moment from his exposition of Copernicus and seeks to establish a connection between the fortunes of earthly monarchies and the motion of the center of the earth's eccentric round the small circle which it describes about its mean position (see Fig. 18, supra). He believed that Rome began to decline when this center started approaching the sun, and that, when at length it has moved round to the position which it occupied at the Creation, the Second Advent may be expected. (Rheticus' astrological leanings may be attributed to the influence of his mentor Melanchthon more probably than to that of Copernicus, whose writings seem to be free from all trace of the false science.)

Returning to his text, Rheticus deals with the remaining heads in the following somewhat arbitrary order: the Copernican lunar theory; the arguments in favor of the mobility of the earth; the general arrangement of the heliocentric solar system (with the diagram of *De rev.*, I, 11); the threefold motion of the earth and the double oscillation of the polar axis; and, lastly, the Copernican schemes of motion of the planets in longitude and latitude, without numerical data or diagrams.

Select Bibliography

(Incidental reference is made to other books and papers in the text.)

A. Texts and Translations

ARISTOTLE. *Works,* translated into English, ed. W. D. Ross. Vol. II (*Physica, De caelo,* etc.). Oxford, 1922–30.

BRAHE, TYCHO. *Opera omnia,* ed. J. L. E. Dreyer. Copenhagen 1913–29. 15 vols.

COPERNICUS, N. *De revolutionibus orbium coelestium libri VI.* Nuremberg, 1543. Later editions based upon the text of this first edition are those of Basle, 1566; Amsterdam, 1617; Warsaw, 1854. The Warsaw edition contains a Polish translation. These editions were superseded by that of M. Curtze (Torun, 1873), which was based upon a critical study of the original manuscript. The new *Gesamtausgabe* of the works of Copernicus (Munich, 1944 etc.) includes a facsimile of the manuscript of the *De revolutionibus* (Band I) and an edition of the text by F. and C. Zeller (Band II). There is a German translation of the text of 1873 by C. L. Menzzer, Torun, 1879; an English translation of the Preface and Book I by J. F. Dobson and S. Brodetsky (Occasional Notes of the Royal Astronomical Society, London, 1947); and English versions of selected passages in H. Shapley and H. E. Howarth, *A Source Book in Astronomy,* New York, 1929. The first eleven chapters of the *De revolutionibus* have been translated into French, with introduction and notes, by A. Koyré, Paris, 1934.

———. *De hypothesibus motuum coelestium a se constitutis commentariolus.* MS. At least two copies of this manuscript are known; upon Curtze's collation of them, Prowe based the text which he published in the second volume of his *Nicolaus Coppernicus,* 184–202. There is an English translation of the *Commentariolus,* based on Prowe's text, by Edward Rosen in *Osiris,* Vol. III, Part 1, 1937; and a revised version of this translation is included in E. Rosen, *Three Copernican Treatises,* translated with introduction and notes, New York, 1939.

DESCARTES, R. *Œuvres,* ed. C. Adam and P. Tannery. Paris, 1897–1910. 12 vols.

GALILEO. *Opere,* ed. A. Favaro. Florence, 1890–1909. 20 vols. The *Dialogo* of 1632 has been translated into German by E. Strauss, Leipzig, 1891, and into English by T. Salusbury in his *Mathematical Collections and Translations,* London, 1661, Vol. I. Salusbury's translation has been revised and edited by Giorgio de Santillana, Chicago, 1953. The *Sidereus nuncius* of 1610 has been translated into English by E. S. Carlos, London, 1880.

HOOKE, R. The relevant material is available in R. T. Gunther, *Early Science in Oxford,* Vols. VI, VII, VIII, XIII. Oxford, 1923 etc.

HUYGENS, C. *Œuvres complètes.* The Hague, 1888–1950, 22 vols.

KEPLER, J. *Opera omnia,* ed. C. Frisch. Frankfurt and Erlangen, 1858–71. 8 vols. A fine modern edition has been in process of publication by Max Caspar and the late Walther von Dyck since 1937. German translations by Max Caspar have appeared of Kepler's *Mysterium cosmographicum* (1923), *Astronomia nova* (1929), and *Harmonice mundi* (1939).

NEWTON, ISAAC. *Philosophiae naturalis principia mathematica.* London, 1687. Andrew Motte's English translation has been revised and annotated by Florian Cajori, Cambridge, 1934.

PLATO. *Timaeus and Critias,* translated into English by A. E. Taylor. London, 1929.

PTOLEMAEUS, C. *Syntaxis mathematica* (the *Almagest*), ed. J. L. Heiberg. Leipzig, 1898–1903. There is a German translation by K. Manitius, Leipzig, 1912–13, and a French translation, based upon an inferior text, by Halma, Paris, 1813–16.

RHETICUS, G. JOACHIMUS. *De libris revolutionum narratio prima.* Danzig, 1540. This work is included in the Torun edition of the *De revolutionibus,* 1873. An English translation will be found in E. Rosen, *Three Copernican Treatises,* New York, 1939.

B. Other Works [1]

BALL, W. W. R. *An Essay on Newton's "Principia."* London, 1893.

BIRKENMAJER, L. A. *Mikolaj Kopernik.* Cracow, 1900. Use has been made in the present volume of the French summary of this work found in *Bulletin International de l'Académie des Sciences de Cracovie, Classe des Sciences Math. et Nat.* March 1902, 200 ff.

―――. *Stromata Copernicana.* Cracow, 1924.

BURTT, E. A. *The Metaphysical Foundations of Modern Physical Science.* (Revised edition.) London, 1932. *

CASPAR, M. *Johannes Kepler.* Stuttgart, 1948. *

---

[1] Of the following list of titles, those marked by an asterisk are currently available from Dover Publications, Inc. Please visit www.doverpublications.com for details.

CROMBIE, A. C. *Augustine to Galileo: The History of Science A.D. 400–1650.* London, 1957.*

DREYER, J. L. E. *History of the Planetary Systems from Thales to Kepler.* Cambridge, 1906. *

——. *Tycho Brahe.* Edinburgh, 1890.

DUHEM, P. *Le Système du monde.* Paris, 1913–17. 5 vols.

GASSENDI, P. *Tychonis Brahei vita; accessit N. Copernici vita.* Paris, 1654.

HEATH, SIR T. L. *Aristarchus of Samos, the Ancient Copernicus.* Oxford, 1913.

HERZ, N. *Geschichte der Bahnbestimmung von Planeten und Cometen.* Leipzig, 1887, 1894.

JOHNSON, F. R. *Astronomical Thought in Renaissance England.* Baltimore, 1937.

KOYRÉ, A. *Études Galiléennes.* Paris, 1939. 3 vols.

KUGLER, F. X. *Sternkunde und Sterndienst in Babel.* Münster, 1907 etc.

NEUGEBAUER, O. *The Exact Sciences in Antiquity.* Princeton, 1951.*

PROWE, L. *Nicolaus Coppernicus.* Berlin, 1883, 1884.

SARTON, G. *Introduction to the History of Science.* Washington, 1927–48. 3 vols.

SINGER, DOROTHEA W. *Giordano Bruno: His Life and Thought.* New York, 1950.

STIMSON, DOROTHY. *The Gradual Acceptance of the Copernican Theory of the Universe.* New York, 1917.

TANNERY, P. *Recherches sur l'histoire de l'astronomie ancienne.* Paris, 1893.

WOHLWILL, E. *Galilei und sein Kampf für die Copernicanische Lehre.* Hamburg, 1909, 1926.

ZINNER, E. *Entstehung und Ausbreitung der Coppernicanischen Lehre.* Erlangen, 1943.

Index

A CATALOG OF SELECTED
DOVER BOOKS
IN ALL FIELDS OF INTEREST

A CATALOG OF SELECTED DOVER
BOOKS IN ALL FIELDS OF INTEREST

CONCERNING THE SPIRITUAL IN ART, Wassily Kandinsky. Pioneering work by father of abstract art. Thoughts on color theory, nature of art. Analysis of earlier masters. 12 illustrations. 80pp. of text. 5⅜ x 8½. 23411-8

ANIMALS: 1,419 Copyright-Free Illustrations of Mammals, Birds, Fish, Insects, etc., Jim Harter (ed.). Clear wood engravings present, in extremely lifelike poses, over 1,000 species of animals. One of the most extensive pictorial sourcebooks of its kind. Captions. Index. 284pp. 9 x 12. 23766-4

CELTIC ART: The Methods of Construction, George Bain. Simple geometric techniques for making Celtic interlacements, spirals, Kells-type initials, animals, humans, etc. Over 500 illustrations. 160pp. 9 x 12. (Available in U.S. only.) 22923-8

AN ATLAS OF ANATOMY FOR ARTISTS, Fritz Schider. Most thorough reference work on art anatomy in the world. Hundreds of illustrations, including selections from works by Vesalius, Leonardo, Goya, Ingres, Michelangelo, others. 593 illustrations. 192pp. 7⅛ x 10¼. 20241-0

CELTIC HAND STROKE-BY-STROKE (Irish Half-Uncial from "The Book of Kells"): An Arthur Baker Calligraphy Manual, Arthur Baker. Complete guide to creating each letter of the alphabet in distinctive Celtic manner. Covers hand position, strokes, pens, inks, paper, more. Illustrated. 48pp. 8¼ x 11. 24336-2

EASY ORIGAMI, John Montroll. Charming collection of 32 projects (hat, cup, pelican, piano, swan, many more) specially designed for the novice origami hobbyist. Clearly illustrated easy-to-follow instructions insure that even beginning papercrafters will achieve successful results. 48pp. 8¼ x 11. 27298-2

THE COMPLETE BOOK OF BIRDHOUSE CONSTRUCTION FOR WOODWORKERS, Scott D. Campbell. Detailed instructions, illustrations, tables. Also data on bird habitat and instinct patterns. Bibliography. 3 tables. 63 illustrations in 15 figures. 48pp. 5¼ x 8½. 24407-5

BLOOMINGDALE'S ILLUSTRATED 1886 CATALOG: Fashions, Dry Goods and Housewares, Bloomingdale Brothers. Famed merchants' extremely rare catalog depicting about 1,700 products: clothing, housewares, firearms, dry goods, jewelry, more. Invaluable for dating, identifying vintage items. Also, copyright-free graphics for artists, designers. Co-published with Henry Ford Museum & Greenfield Village. 160pp. 8¼ x 11. 25780-0

HISTORIC COSTUME IN PICTURES, Braun & Schneider. Over 1,450 costumed figures in clearly detailed engravings—from dawn of civilization to end of 19th century. Captions. Many folk costumes. 256pp. 8⅜ x 11¾. 23150-X

STICKLEY CRAFTSMAN FURNITURE CATALOGS, Gustav Stickley and L. & J. G. Stickley. Beautiful, functional furniture in two authentic catalogs from 1910. 594 illustrations, including 277 photos, show settles, rockers, armchairs, reclining chairs, bookcases, desks, tables. 183pp. 6½ x 9¼. 23838-5

AMERICAN LOCOMOTIVES IN HISTORIC PHOTOGRAPHS: 1858 to 1949, Ron Ziel (ed.). A rare collection of 126 meticulously detailed official photographs, called "builder portraits," of American locomotives that majestically chronicle the rise of steam locomotive power in America. Introduction. Detailed captions. xi+129pp. 9 x 12. 27393-8

AMERICA'S LIGHTHOUSES: An Illustrated History, Francis Ross Holland, Jr. Delightfully written, profusely illustrated fact-filled survey of over 200 American lighthouses since 1716. History, anecdotes, technological advances, more. 240pp. 8 x 10¾. 25576-X

TOWARDS A NEW ARCHITECTURE, Le Corbusier. Pioneering manifesto by founder of "International School." Technical and aesthetic theories, views of industry, economics, relation of form to function, "mass-production split" and much more. Profusely illustrated. 320pp. 6⅛ x 9¼. (Available in U.S. only.) 25023-7

HOW THE OTHER HALF LIVES, Jacob Riis. Famous journalistic record, exposing poverty and degradation of New York slums around 1900, by major social reformer. 100 striking and influential photographs. 233pp. 10 x 7⅞. 22012-5

FRUIT KEY AND TWIG KEY TO TREES AND SHRUBS, William M. Harlow. One of the handiest and most widely used identification aids. Fruit key covers 120 deciduous and evergreen species; twig key 160 deciduous species. Easily used. Over 300 photographs. 126pp. 5⅜ x 8½. 20511-8

COMMON BIRD SONGS, Dr. Donald J. Borror. Songs of 60 most common U.S. birds: robins, sparrows, cardinals, bluejays, finches, more—arranged in order of increasing complexity. Up to 9 variations of songs of each species.
Cassette and manual 99911-4

ORCHIDS AS HOUSE PLANTS, Rebecca Tyson Northen. Grow cattleyas and many other kinds of orchids—in a window, in a case, or under artificial light. 63 illustrations. 148pp. 5⅜ x 8½. 23261-1

MONSTER MAZES, Dave Phillips. Masterful mazes at four levels of difficulty. Avoid deadly perils and evil creatures to find magical treasures. Solutions for all 32 exciting illustrated puzzles. 48pp. 8¼ x 11. 26005-4

MOZART'S DON GIOVANNI (DOVER OPERA LIBRETTO SERIES), Wolfgang Amadeus Mozart. Introduced and translated by Ellen H. Bleiler. Standard Italian libretto, with complete English translation. Convenient and thoroughly portable—an ideal companion for reading along with a recording or the performance itself. Introduction. List of characters. Plot summary. 121pp. 5¼ x 8½. 24944-1

TECHNICAL MANUAL AND DICTIONARY OF CLASSICAL BALLET, Gail Grant. Defines, explains, comments on steps, movements, poses and concepts. 15-page pictorial section. Basic book for student, viewer. 127pp. 5⅜ x 8½. 21843-0

THE CLARINET AND CLARINET PLAYING, David Pino. Lively, comprehensive work features suggestions about technique, musicianship, and musical interpretation, as well as guidelines for teaching, making your own reeds, and preparing for public performance. Includes an intriguing look at clarinet history. "A godsend," *The Clarinet*, Journal of the International Clarinet Society. Appendixes. 7 illus. 320pp. 5⅜ x 8½. 40270-3

HOLLYWOOD GLAMOR PORTRAITS, John Kobal (ed.). 145 photos from 1926-49. Harlow, Gable, Bogart, Bacall; 94 stars in all. Full background on photographers, technical aspects. 160pp. 8⅜ x 11¼. 23352-9

THE ANNOTATED CASEY AT THE BAT: A Collection of Ballads about the Mighty Casey/Third, Revised Edition, Martin Gardner (ed.). Amusing sequels and parodies of one of America's best-loved poems: Casey's Revenge, Why Casey Whiffed, Casey's Sister at the Bat, others. 256pp. 5⅜ x 8½. 28598-7

THE RAVEN AND OTHER FAVORITE POEMS, Edgar Allan Poe. Over 40 of the author's most memorable poems: "The Bells," "Ulalume," "Israfel," "To Helen," "The Conqueror Worm," "Eldorado," "Annabel Lee," many more. Alphabetic lists of titles and first lines. 64pp. 5³⁄₁₆ x 8¼. 26685-0

PERSONAL MEMOIRS OF U. S. GRANT, Ulysses Simpson Grant. Intelligent, deeply moving firsthand account of Civil War campaigns, considered by many the finest military memoirs ever written. Includes letters, historic photographs, maps and more. 528pp. 6⅛ x 9¼. 28587-1

ANCIENT EGYPTIAN MATERIALS AND INDUSTRIES, A. Lucas and J. Harris. Fascinating, comprehensive, thoroughly documented text describes this ancient civilization's vast resources and the processes that incorporated them in daily life, including the use of animal products, building materials, cosmetics, perfumes and incense, fibers, glazed ware, glass and its manufacture, materials used in the mummification process, and much more. 544pp. 6⅛ x 9¼. (Available in U.S. only.) 40446-3

RUSSIAN STORIES/RUSSKIE RASSKAZY: A Dual-Language Book, edited by Gleb Struve. Twelve tales by such masters as Chekhov, Tolstoy, Dostoevsky, Pushkin, others. Excellent word-for-word English translations on facing pages, plus teaching and study aids, Russian/English vocabulary, biographical/critical introductions, more. 416pp. 5⅜ x 8½. 26244-8

PHILADELPHIA THEN AND NOW: 60 Sites Photographed in the Past and Present, Kenneth Finkel and Susan Oyama. Rare photographs of City Hall, Logan Square, Independence Hall, Betsy Ross House, other landmarks juxtaposed with contemporary views. Captures changing face of historic city. Introduction. Captions. 128pp. 8¼ x 11. 25790-8

AIA ARCHITECTURAL GUIDE TO NASSAU AND SUFFOLK COUNTIES, LONG ISLAND, The American Institute of Architects, Long Island Chapter, and the Society for the Preservation of Long Island Antiquities. Comprehensive, well-researched and generously illustrated volume brings to life over three centuries of Long Island's great architectural heritage. More than 240 photographs with authoritative, extensively detailed captions. 176pp. 8¼ x 11. 26946-9

NORTH AMERICAN INDIAN LIFE: Customs and Traditions of 23 Tribes, Elsie Clews Parsons (ed.). 27 fictionalized essays by noted anthropologists examine religion, customs, government, additional facets of life among the Winnebago, Crow, Zuni, Eskimo, other tribes. 480pp. 6⅛ x 9¼. 27377-6

FRANK LLOYD WRIGHT'S DANA HOUSE, Donald Hoffmann. Pictorial essay of residential masterpiece with over 160 interior and exterior photos, plans, elevations, sketches and studies. 128pp. 9¼ x 10¾. 29120-0

THE MALE AND FEMALE FIGURE IN MOTION: 60 Classic Photographic Sequences, Eadweard Muybridge. 60 true-action photographs of men and women walking, running, climbing, bending, turning, etc., reproduced from rare 19th-century masterpiece. vi + 121pp. 9 x 12. 24745-7

1001 QUESTIONS ANSWERED ABOUT THE SEASHORE, N. J. Berrill and Jacquelyn Berrill. Queries answered about dolphins, sea snails, sponges, starfish, fishes, shore birds, many others. Covers appearance, breeding, growth, feeding, much more. 305pp. 5¼ x 8¼. 23366-9

ATTRACTING BIRDS TO YOUR YARD, William J. Weber. Easy-to-follow guide offers advice on how to attract the greatest diversity of birds: birdhouses, feeders, water and waterers, much more. 96pp. 5³⁄₁₆ x 8¼. 28927-3

MEDICINAL AND OTHER USES OF NORTH AMERICAN PLANTS: A Historical Survey with Special Reference to the Eastern Indian Tribes, Charlotte Erichsen-Brown. Chronological historical citations document 500 years of usage of plants, trees, shrubs native to eastern Canada, northeastern U.S. Also complete identifying information. 343 illustrations. 544pp. 6½ x 9¼. 25951-X

STORYBOOK MAZES, Dave Phillips. 23 stories and mazes on two-page spreads: Wizard of Oz, Treasure Island, Robin Hood, etc. Solutions. 64pp. 8¼ x 11. 23628-5

AMERICAN NEGRO SONGS: 230 Folk Songs and Spirituals, Religious and Secular, John W. Work. This authoritative study traces the African influences of songs sung and played by black Americans at work, in church, and as entertainment. The author discusses the lyric significance of such songs as "Swing Low, Sweet Chariot," "John Henry," and others and offers the words and music for 230 songs. Bibliography. Index of Song Titles. 272pp. 6½ x 9¼. 40271-1

MOVIE-STAR PORTRAITS OF THE FORTIES, John Kobal (ed.). 163 glamor, studio photos of 106 stars of the 1940s: Rita Hayworth, Ava Gardner, Marlon Brando, Clark Gable, many more. 176pp. 8⅜ x 11¼. 23546-7

BENCHLEY LOST AND FOUND, Robert Benchley. Finest humor from early 30s, about pet peeves, child psychologists, post office and others. Mostly unavailable elsewhere. 73 illustrations by Peter Arno and others. 183pp. 5⅜ x 8½. 22410-4

YEKL and THE IMPORTED BRIDEGROOM AND OTHER STORIES OF YIDDISH NEW YORK, Abraham Cahan. Film Hester Street based on *Yekl* (1896). Novel, other stories among first about Jewish immigrants on N.Y.'s East Side. 240pp. 5⅜ x 8½. 22427-9

SELECTED POEMS, Walt Whitman. Generous sampling from *Leaves of Grass*. Twenty-four poems include "I Hear America Singing," "Song of the Open Road," "I Sing the Body Electric," "When Lilacs Last in the Dooryard Bloom'd," "O Captain! My Captain!"—all reprinted from an authoritative edition. Lists of titles and first lines. 128pp. 5³⁄₁₆ x 8¼. 26878-0

THE BEST TALES OF HOFFMANN, E. T. A. Hoffmann. 10 of Hoffmann's most important stories: "Nutcracker and the King of Mice," "The Golden Flowerpot," etc. 458pp. 5⅜ x 8½. 21793-0

FROM FETISH TO GOD IN ANCIENT EGYPT, E. A. Wallis Budge. Rich detailed survey of Egyptian conception of "God" and gods, magic, cult of animals, Osiris, more. Also, superb English translations of hymns and legends. 240 illustrations. 545pp. 5⅜ x 8½. 25803-3

FRENCH STORIES/CONTES FRANÇAIS: A Dual-Language Book, Wallace Fowlie. Ten stories by French masters, Voltaire to Camus: "Micromegas" by Voltaire; "The Atheist's Mass" by Balzac; "Minuet" by de Maupassant; "The Guest" by Camus, six more. Excellent English translations on facing pages. Also French-English vocabulary list, exercises, more. 352pp. 5⅜ x 8½. 26443-2

CHICAGO AT THE TURN OF THE CENTURY IN PHOTOGRAPHS: 122 Historic Views from the Collections of the Chicago Historical Society, Larry A. Viskochil. Rare large-format prints offer detailed views of City Hall, State Street, the Loop, Hull House, Union Station, many other landmarks, circa 1904-1913. Introduction. Captions. Maps. 144pp. 9⅜ x 12¼. 24656-6

OLD BROOKLYN IN EARLY PHOTOGRAPHS, 1865-1929, William Lee Younger. Luna Park, Gravesend race track, construction of Grand Army Plaza, moving of Hotel Brighton, etc. 157 previously unpublished photographs. 165pp. 8⅞ x 11¾. 23587-4

THE MYTHS OF THE NORTH AMERICAN INDIANS, Lewis Spence. Rich anthology of the myths and legends of the Algonquins, Iroquois, Pawnees and Sioux, prefaced by an extensive historical and ethnological commentary. 36 illustrations. 480pp. 5⅜ x 8½. 25967-6

AN ENCYCLOPEDIA OF BATTLES: Accounts of Over 1,560 Battles from 1479 B.C. to the Present, David Eggenberger. Essential details of every major battle in recorded history from the first battle of Megiddo in 1479 B.C. to Grenada in 1984. List of Battle Maps. New Appendix covering the years 1967-1984. Index. 99 illustrations. 544pp. 6½ x 9¼. 24913-1

SAILING ALONE AROUND THE WORLD, Captain Joshua Slocum. First man to sail around the world, alone, in small boat. One of great feats of seamanship told in delightful manner. 67 illustrations. 294pp. 5⅜ x 8½. 20326-3

ANARCHISM AND OTHER ESSAYS, Emma Goldman. Powerful, penetrating, prophetic essays on direct action, role of minorities, prison reform, puritan hypocrisy, violence, etc. 271pp. 5⅜ x 8½. 22484-8

MYTHS OF THE HINDUS AND BUDDHISTS, Ananda K. Coomaraswamy and Sister Nivedita. Great stories of the epics; deeds of Krishna, Shiva, taken from puranas, Vedas, folk tales; etc. 32 illustrations. 400pp. 5⅜ x 8½. 21759-0

THE TRAUMA OF BIRTH, Otto Rank. Rank's controversial thesis that anxiety neurosis is caused by profound psychological trauma which occurs at birth. 256pp. 5⅜ x 8½. 27974-X

A THEOLOGICO-POLITICAL TREATISE, Benedict Spinoza. Also contains unfinished Political Treatise. Great classic on religious liberty, theory of government on common consent. R. Elwes translation. Total of 421pp. 5⅜ x 8½. 20249-6

MY BONDAGE AND MY FREEDOM, Frederick Douglass. Born a slave, Douglass became outspoken force in antislavery movement. The best of Douglass' autobiographies. Graphic description of slave life. 464pp. 5⅜ x 8½. 22457-0

FOLLOWING THE EQUATOR: A Journey Around the World, Mark Twain. Fascinating humorous account of 1897 voyage to Hawaii, Australia, India, New Zealand, etc. Ironic, bemused reports on peoples, customs, climate, flora and fauna, politics, much more. 197 illustrations. 720pp. 5⅜ x 8½. 26113-1

THE PEOPLE CALLED SHAKERS, Edward D. Andrews. Definitive study of Shakers: origins, beliefs, practices, dances, social organization, furniture and crafts, etc. 33 illustrations. 351pp. 5⅜ x 8½. 21081-2

THE MYTHS OF GREECE AND ROME, H. A. Guerber. A classic of mythology, generously illustrated, long prized for its simple, graphic, accurate retelling of the principal myths of Greece and Rome, and for its commentary on their origins and significance. With 64 illustrations by Michelangelo, Raphael, Titian, Rubens, Canova, Bernini and others. 480pp. 5⅜ x 8½. 27584-1

PSYCHOLOGY OF MUSIC, Carl E. Seashore. Classic work discusses music as a medium from psychological viewpoint. Clear treatment of physical acoustics, auditory apparatus, sound perception, development of musical skills, nature of musical feeling, host of other topics. 88 figures. 408pp. 5⅜ x 8½. 21851-1

THE PHILOSOPHY OF HISTORY, Georg W. Hegel. Great classic of Western thought develops concept that history is not chance but rational process, the evolution of freedom. 457pp. 5⅜ x 8½. 20112-0

THE BOOK OF TEA, Kakuzo Okakura. Minor classic of the Orient: entertaining, charming explanation, interpretation of traditional Japanese culture in terms of tea ceremony. 94pp. 5⅜ x 8½. 20070-1

LIFE IN ANCIENT EGYPT, Adolf Erman. Fullest, most thorough, detailed older account with much not in more recent books, domestic life, religion, magic, medicine, commerce, much more. Many illustrations reproduce tomb paintings, carvings, hieroglyphs, etc. 597pp. 5⅜ x 8½. 22632-8

SUNDIALS, Their Theory and Construction, Albert Waugh. Far and away the best, most thorough coverage of ideas, mathematics concerned, types, construction, adjusting anywhere. Simple, nontechnical treatment allows even children to build several of these dials. Over 100 illustrations. 230pp. 5⅜ x 8½. 22947-5

THEORETICAL HYDRODYNAMICS, L. M. Milne-Thomson. Classic exposition of the mathematical theory of fluid motion, applicable to both hydrodynamics and aerodynamics. Over 600 exercises. 768pp. 6⅛ x 9¼. 68970-0

SONGS OF EXPERIENCE: Facsimile Reproduction with 26 Plates in Full Color, William Blake. 26 full-color plates from a rare 1826 edition. Includes "The Tyger," "London," "Holy Thursday," and other poems. Printed text of poems. 48pp. 5¼ x 7. 24636-1

OLD-TIME VIGNETTES IN FULL COLOR, Carol Belanger Grafton (ed.). Over 390 charming, often sentimental illustrations, selected from archives of Victorian graphics—pretty women posing, children playing, food, flowers, kittens and puppies, smiling cherubs, birds and butterflies, much more. All copyright-free. 48pp. 9¼ x 12¼. 27269-9

PERSPECTIVE FOR ARTISTS, Rex Vicat Cole. Depth, perspective of sky and sea, shadows, much more, not usually covered. 391 diagrams, 81 reproductions of drawings and paintings. 279pp. 5⅜ x 8½. 22487-2

DRAWING THE LIVING FIGURE, Joseph Sheppard. Innovative approach to artistic anatomy focuses on specifics of surface anatomy, rather than muscles and bones. Over 170 drawings of live models in front, back and side views, and in widely varying poses. Accompanying diagrams. 177 illustrations. Introduction. Index. 144pp. 8⅜ x11¼. 26723-7

GOTHIC AND OLD ENGLISH ALPHABETS: 100 Complete Fonts, Dan X. Solo. Add power, elegance to posters, signs, other graphics with 100 stunning copyright-free alphabets: Blackstone, Dolbey, Germania, 97 more–including many lower-case, numerals, punctuation marks. 104pp. 8⅛ x 11. 24695-7

HOW TO DO BEADWORK, Mary White. Fundamental book on craft from simple projects to five-bead chains and woven works. 106 illustrations. 142pp. 5⅜ x 8. 20697-1

THE BOOK OF WOOD CARVING, Charles Marshall Sayers. Finest book for beginners discusses fundamentals and offers 34 designs. "Absolutely first rate . . . well thought out and well executed."–E. J. Tangerman. 118pp. 7¾ x 10⅝. 23654-4

ILLUSTRATED CATALOG OF CIVIL WAR MILITARY GOODS: Union Army Weapons, Insignia, Uniform Accessories, and Other Equipment, Schuyler, Hartley, and Graham. Rare, profusely illustrated 1846 catalog includes Union Army uniform and dress regulations, arms and ammunition, coats, insignia, flags, swords, rifles, etc. 226 illustrations. 160pp. 9 x 12. 24939-5

WOMEN'S FASHIONS OF THE EARLY 1900s: An Unabridged Republication of "New York Fashions, 1909," National Cloak & Suit Co. Rare catalog of mail-order fashions documents women's and children's clothing styles shortly after the turn of the century. Captions offer full descriptions, prices. Invaluable resource for fashion, costume historians. Approximately 725 illustrations. 128pp. 8⅜ x 11¼. 27276-1

THE 1912 AND 1915 GUSTAV STICKLEY FURNITURE CATALOGS, Gustav Stickley. With over 200 detailed illustrations and descriptions, these two catalogs are essential reading and reference materials and identification guides for Stickley furniture. Captions cite materials, dimensions and prices. 112pp. 6½ x 9¼. 26676-1

EARLY AMERICAN LOCOMOTIVES, John H. White, Jr. Finest locomotive engravings from early 19th century: historical (1804–74), main-line (after 1870), special, foreign, etc. 147 plates. 142pp. 11⅜ x 8¼. 22772-3

THE TALL SHIPS OF TODAY IN PHOTOGRAPHS, Frank O. Braynard. Lavishly illustrated tribute to nearly 100 majestic contemporary sailing vessels: Amerigo Vespucci, Clearwater, Constitution, Eagle, Mayflower, Sea Cloud, Victory, many more. Authoritative captions provide statistics, background on each ship. 190 black-and-white photographs and illustrations. Introduction. 128pp. 8⅞ x 11¾. 27163-3

LITTLE BOOK OF EARLY AMERICAN CRAFTS AND TRADES, Peter Stockham (ed.). 1807 children's book explains crafts and trades: baker, hatter, cooper, potter, and many others. 23 copperplate illustrations. 140pp. 4⅝ x 6. 23336-7

VICTORIAN FASHIONS AND COSTUMES FROM HARPER'S BAZAR, 1867–1898, Stella Blum (ed.). Day costumes, evening wear, sports clothes, shoes, hats, other accessories in over 1,000 detailed engravings. 320pp. 9⅜ x 12¼. 22990-4

GUSTAV STICKLEY, THE CRAFTSMAN, Mary Ann Smith. Superb study surveys broad scope of Stickley's achievement, especially in architecture. Design philosophy, rise and fall of the Craftsman empire, descriptions and floor plans for many Craftsman houses, more. 86 black-and-white halftones. 31 line illustrations. Introduction 208pp. 6½ x 9¼. 27210-9

THE LONG ISLAND RAIL ROAD IN EARLY PHOTOGRAPHS, Ron Ziel. Over 220 rare photos, informative text document origin (1844) and development of rail service on Long Island. Vintage views of early trains, locomotives, stations, passengers, crews, much more. Captions. 8⅞ x 11¾. 26301-0

VOYAGE OF THE LIBERDADE, Joshua Slocum. Great 19th-century mariner's thrilling, first-hand account of the wreck of his ship off South America, the 35-foot boat he built from the wreckage, and its remarkable voyage home. 128pp. 5⅜ x 8½. 40022-0

TEN BOOKS ON ARCHITECTURE, Vitruvius. The most important book ever written on architecture. Early Roman aesthetics, technology, classical orders, site selection, all other aspects. Morgan translation. 331pp. 5⅜ x 8½. 20645-9

THE HUMAN FIGURE IN MOTION, Eadweard Muybridge. More than 4,500 stopped-action photos, in action series, showing undraped men, women, children jumping, lying down, throwing, sitting, wrestling, carrying, etc. 390pp. 7⅞ x 10⅝. 20204-6 Clothbd.

TREES OF THE EASTERN AND CENTRAL UNITED STATES AND CANADA, William M. Harlow. Best one-volume guide to 140 trees. Full descriptions, woodlore, range, etc. Over 600 illustrations. Handy size. 288pp. 4½ x 6⅜. 20395-6

SONGS OF WESTERN BIRDS, Dr. Donald J. Borror. Complete song and call repertoire of 60 western species, including flycatchers, juncoos, cactus wrens, many more–includes fully illustrated booklet. Cassette and manual 99913-0

GROWING AND USING HERBS AND SPICES, Milo Miloradovich. Versatile handbook provides all the information needed for cultivation and use of all the herbs and spices available in North America. 4 illustrations. Index. Glossary. 236pp. 5⅜ x 8½. 25058-X

BIG BOOK OF MAZES AND LABYRINTHS, Walter Shepherd. 50 mazes and labyrinths in all–classical, solid, ripple, and more–in one great volume. Perfect inexpensive puzzler for clever youngsters. Full solutions. 112pp. 8⅛ x 11. 22951-3

PIANO TUNING, J. Cree Fischer. Clearest, best book for beginner, amateur. Simple repairs, raising dropped notes, tuning by easy method of flattened fifths. No previous skills needed. 4 illustrations. 201pp. 5⅜ x 8½. 23267-0

HINTS TO SINGERS, Lillian Nordica. Selecting the right teacher, developing confidence, overcoming stage fright, and many other important skills receive thoughtful discussion in this indispensible guide, written by a world-famous diva of four decades' experience. 96pp. 5⅜ x 8½. 40094-8

THE COMPLETE NONSENSE OF EDWARD LEAR, Edward Lear. All nonsense limericks, zany alphabets, Owl and Pussycat, songs, nonsense botany, etc., illustrated by Lear. Total of 320pp. 5⅜ x 8½. (Available in U.S. only.) 20167-8

VICTORIAN PARLOUR POETRY: An Annotated Anthology, Michael R. Turner. 117 gems by Longfellow, Tennyson, Browning, many lesser-known poets. "The Village Blacksmith," "Curfew Must Not Ring Tonight," "Only a Baby Small," dozens more, often difficult to find elsewhere. Index of poets, titles, first lines. xxiii + 325pp. 5⅝ x 8¼. 27044-0

DUBLINERS, James Joyce. Fifteen stories offer vivid, tightly focused observations of the lives of Dublin's poorer classes. At least one, "The Dead," is considered a masterpiece. Reprinted complete and unabridged from standard edition. 160pp. $5\frac{3}{16}$ x 8¼. 26870-5

GREAT WEIRD TALES: 14 Stories by Lovecraft, Blackwood, Machen and Others, S. T. Joshi (ed.). 14 spellbinding tales, including "The Sin Eater," by Fiona McLeod, "The Eye Above the Mantel," by Frank Belknap Long, as well as renowned works by R. H. Barlow, Lord Dunsany, Arthur Machen, W. C. Morrow and eight other masters of the genre. 256pp. 5⅜ x 8½. (Available in U.S. only.) 40436-6

THE BOOK OF THE SACRED MAGIC OF ABRAMELIN THE MAGE, translated by S. MacGregor Mathers. Medieval manuscript of ceremonial magic. Basic document in Aleister Crowley, Golden Dawn groups. 268pp. 5⅜ x 8½. 23211-5

NEW RUSSIAN-ENGLISH AND ENGLISH-RUSSIAN DICTIONARY, M. A. O'Brien. This is a remarkably handy Russian dictionary, containing a surprising amount of information, including over 70,000 entries. 366pp. 4½ x 6⅛. 20208-9

HISTORIC HOMES OF THE AMERICAN PRESIDENTS, Second, Revised Edition, Irvin Haas. A traveler's guide to American Presidential homes, most open to the public, depicting and describing homes occupied by every American President from George Washington to George Bush. With visiting hours, admission charges, travel routes. 175 photographs. Index. 160pp. 8¼ x 11. 26751-2

NEW YORK IN THE FORTIES, Andreas Feininger. 162 brilliant photographs by the well-known photographer, formerly with *Life* magazine. Commuters, shoppers, Times Square at night, much else from city at its peak. Captions by John von Hartz. 181pp. 9¼ x 10¾. 23585-8

INDIAN SIGN LANGUAGE, William Tomkins. Over 525 signs developed by Sioux and other tribes. Written instructions and diagrams. Also 290 pictographs. 111pp. 6⅛ x 9¼. 22029-X

ANATOMY: A Complete Guide for Artists, Joseph Sheppard. A master of figure drawing shows artists how to render human anatomy convincingly. Over 460 illustrations. 224pp. 8⅜ x 11¼. 27279-6

MEDIEVAL CALLIGRAPHY: Its History and Technique, Marc Drogin. Spirited history, comprehensive instruction manual covers 13 styles (ca. 4th century through 15th). Excellent photographs; directions for duplicating medieval techniques with modern tools. 224pp. 8⅜ x 11¼. 26142-5

DRIED FLOWERS: How to Prepare Them, Sarah Whitlock and Martha Rankin. Complete instructions on how to use silica gel, meal and borax, perlite aggregate, sand and borax, glycerine and water to create attractive permanent flower arrangements. 12 illustrations. 32pp. 5⅜ x 8½. 21802-3

EASY-TO-MAKE BIRD FEEDERS FOR WOODWORKERS, Scott D. Campbell. Detailed, simple-to-use guide for designing, constructing, caring for and using feeders. Text, illustrations for 12 classic and contemporary designs. 96pp. 5⅜ x 8½.
25847-5

SCOTTISH WONDER TALES FROM MYTH AND LEGEND, Donald A. Mackenzie. 16 lively tales tell of giants rumbling down mountainsides, of a magic wand that turns stone pillars into warriors, of gods and goddesses, evil hags, powerful forces and more. 240pp. 5⅜ x 8½. 29677-6

THE HISTORY OF UNDERCLOTHES, C. Willett Cunnington and Phyllis Cunnington. Fascinating, well-documented survey covering six centuries of English undergarments, enhanced with over 100 illustrations: 12th-century laced-up bodice, footed long drawers (1795), 19th-century bustles, 19th-century corsets for men, Victorian "bust improvers," much more. 272pp. 5⅜ x 8¼. 27124-2

ARTS AND CRAFTS FURNITURE: The Complete Brooks Catalog of 1912, Brooks Manufacturing Co. Photos and detailed descriptions of more than 150 now very collectible furniture designs from the Arts and Crafts movement depict davenports, settees, buffets, desks, tables, chairs, bedsteads, dressers and more, all built of solid, quarter-sawed oak. Invaluable for students and enthusiasts of antiques, Americana and the decorative arts. 80pp. 6½ x 9¼. 27471-3

WILBUR AND ORVILLE: A Biography of the Wright Brothers, Fred Howard. Definitive, crisply written study tells the full story of the brothers' lives and work. A vividly written biography, unparalleled in scope and color, that also captures the spirit of an extraordinary era. 560pp. 6⅛ x 9¼. 40297-5

THE ARTS OF THE SAILOR: Knotting, Splicing and Ropework, Hervey Garrett Smith. Indispensable shipboard reference covers tools, basic knots and useful hitches; handsewing and canvas work, more. Over 100 illustrations. Delightful reading for sea lovers. 256pp. 5⅜ x 8½. 26440-8

FRANK LLOYD WRIGHT'S FALLINGWATER: The House and Its History, Second, Revised Edition, Donald Hoffmann. A total revision–both in text and illustrations–of the standard document on Fallingwater, the boldest, most personal architectural statement of Wright's mature years, updated with valuable new material from the recently opened Frank Lloyd Wright Archives. "Fascinating"–*The New York Times*. 116 illustrations. 128pp. 9¼ x 10¾. 27430-6

PHOTOGRAPHIC SKETCHBOOK OF THE CIVIL WAR, Alexander Gardner. 100 photos taken on field during the Civil War. Famous shots of Manassas Harper's Ferry, Lincoln, Richmond, slave pens, etc. 244pp. 10⅝ x 8¼. 22731-6

FIVE ACRES AND INDEPENDENCE, Maurice G. Kains. Great back-to-the-land classic explains basics of self-sufficient farming. The one book to get. 95 illustrations. 397pp. 5⅜ x 8½. 20974-1

SONGS OF EASTERN BIRDS, Dr. Donald J. Borror. Songs and calls of 60 species most common to eastern U.S.: warblers, woodpeckers, flycatchers, thrushes, larks, many more in high-quality recording. Cassette and manual 99912-2

A MODERN HERBAL, Margaret Grieve. Much the fullest, most exact, most useful compilation of herbal material. Gigantic alphabetical encyclopedia, from aconite to zedoary, gives botanical information, medical properties, folklore, economic uses, much else. Indispensable to serious reader. 161 illustrations. 888pp. 6½ x 9¼. 2-vol. set. (Available in U.S. only.) Vol. I: 22798-7 Vol. II: 22799-5

HIDDEN TREASURE MAZE BOOK, Dave Phillips. Solve 34 challenging mazes accompanied by heroic tales of adventure. Evil dragons, people-eating plants, blood-thirsty giants, many more dangerous adversaries lurk at every twist and turn. 34 mazes, stories, solutions. 48pp. 8¼ x 11. 24566-7

LETTERS OF W. A. MOZART, Wolfgang A. Mozart. Remarkable letters show bawdy wit, humor, imagination, musical insights, contemporary musical world; includes some letters from Leopold Mozart. 276pp. 5⅜ x 8½. 22859-2

BASIC PRINCIPLES OF CLASSICAL BALLET, Agrippina Vaganova. Great Russian theoretician, teacher explains methods for teaching classical ballet. 118 illustrations. 175pp. 5⅜ x 8½. 22036-2

THE JUMPING FROG, Mark Twain. Revenge edition. The original story of The Celebrated Jumping Frog of Calaveras County, a hapless French translation, and Twain's hilarious "retranslation" from the French. 12 illustrations. 66pp. 5⅜ x 8½. 22686-7

BEST REMEMBERED POEMS, Martin Gardner (ed.). The 126 poems in this superb collection of 19th- and 20th-century British and American verse range from Shelley's "To a Skylark" to the impassioned "Renascence" of Edna St. Vincent Millay and to Edward Lear's whimsical "The Owl and the Pussycat." 224pp. 5⅜ x 8½. 27165-X

COMPLETE SONNETS, William Shakespeare. Over 150 exquisite poems deal with love, friendship, the tyranny of time, beauty's evanescence, death and other themes in language of remarkable power, precision and beauty. Glossary of archaic terms. 80pp. 5³⁄₁₆ x 8¼. 26686-9

THE BATTLES THAT CHANGED HISTORY, Fletcher Pratt. Eminent historian profiles 16 crucial conflicts, ancient to modern, that changed the course of civilization. 352pp. 5⅜ x 8½. 41129-X

THE WIT AND HUMOR OF OSCAR WILDE, Alvin Redman (ed.). More than 1,000 ripostes, paradoxes, wisecracks: Work is the curse of the drinking classes; I can resist everything except temptation; etc. 258pp. 5⅜ x 8½. 20602-5

SHAKESPEARE LEXICON AND QUOTATION DICTIONARY, Alexander Schmidt. Full definitions, locations, shades of meaning in every word in plays and poems. More than 50,000 exact quotations. 1,485pp. 6½ x 9¼. 2-vol. set.
Vol. 1: 22726-X
Vol. 2: 22727-8

SELECTED POEMS, Emily Dickinson. Over 100 best-known, best-loved poems by one of America's foremost poets, reprinted from authoritative early editions. No comparable edition at this price. Index of first lines. 64pp. 5³⁄₁₆ x 8¼. 26466-1

THE INSIDIOUS DR. FU-MANCHU, Sax Rohmer. The first of the popular mystery series introduces a pair of English detectives to their archnemesis, the diabolical Dr. Fu-Manchu. Flavorful atmosphere, fast-paced action, and colorful characters enliven this classic of the genre. 208pp. 5³⁄₁₆ x 8¼. 29898-1

THE MALLEUS MALEFICARUM OF KRAMER AND SPRENGER, translated by Montague Summers. Full text of most important witchhunter's "bible," used by both Catholics and Protestants. 278pp. 6⅝ x 10. 22802-9

SPANISH STORIES/CUENTOS ESPAÑOLES: A Dual-Language Book, Angel Flores (ed.). Unique format offers 13 great stories in Spanish by Cervantes, Borges, others. Faithful English translations on facing pages. 352pp. 5⅜ x 8½. 25399-6

GARDEN CITY, LONG ISLAND, IN EARLY PHOTOGRAPHS, 1869–1919, Mildred H. Smith. Handsome treasury of 118 vintage pictures, accompanied by carefully researched captions, document the Garden City Hotel fire (1899), the Vanderbilt Cup Race (1908), the first airmail flight departing from the Nassau Boulevard Aerodrome (1911), and much more. 96pp. 8⅞ x 11¾. 40669-5

OLD QUEENS, N.Y., IN EARLY PHOTOGRAPHS, Vincent F. Seyfried and William Asadorian. Over 160 rare photographs of Maspeth, Jamaica, Jackson Heights, and other areas. Vintage views of DeWitt Clinton mansion, 1939 World's Fair and more. Captions. 192pp. 8⅞ x 11. 26358-4

CAPTURED BY THE INDIANS: 15 Firsthand Accounts, 1750-1870, Frederick Drimmer. Astounding true historical accounts of grisly torture, bloody conflicts, relentless pursuits, miraculous escapes and more, by people who lived to tell the tale. 384pp. 5⅜ x 8½. 24901-8

THE WORLD'S GREAT SPEECHES (Fourth Enlarged Edition), Lewis Copeland, Lawrence W. Lamm, and Stephen J. McKenna. Nearly 300 speeches provide public speakers with a wealth of updated quotes and inspiration–from Pericles' funeral oration and William Jennings Bryan's "Cross of Gold Speech" to Malcolm X's powerful words on the Black Revolution and Earl of Spenser's tribute to his sister, Diana, Princess of Wales. 944pp. 5⅜ x 8⅜. 40903-1

THE BOOK OF THE SWORD, Sir Richard F. Burton. Great Victorian scholar/adventurer's eloquent, erudite history of the "queen of weapons"–from prehistory to early Roman Empire. Evolution and development of early swords, variations (sabre, broadsword, cutlass, scimitar, etc.), much more. 336pp. 6⅛ x 9¼.
25434-8

AUTOBIOGRAPHY: The Story of My Experiments with Truth, Mohandas K. Gandhi. Boyhood, legal studies, purification, the growth of the Satyagraha (nonviolent protest) movement. Critical, inspiring work of the man responsible for the freedom of India. 480pp. 5⅜ x 8½. (Available in U.S. only.) 24593-4

CELTIC MYTHS AND LEGENDS, T. W. Rolleston. Masterful retelling of Irish and Welsh stories and tales. Cuchulain, King Arthur, Deirdre, the Grail, many more. First paperback edition. 58 full-page illustrations. 512pp. 5⅜ x 8½. 26507-2

THE PRINCIPLES OF PSYCHOLOGY, William James. Famous long course complete, unabridged. Stream of thought, time perception, memory, experimental methods; great work decades ahead of its time. 94 figures. 1,391pp. 5⅜ x 8½. 2-vol. set.
Vol. I: 20381-6 Vol. II: 20382-4

THE WORLD AS WILL AND REPRESENTATION, Arthur Schopenhauer. Definitive English translation of Schopenhauer's life work, correcting more than 1,000 errors, omissions in earlier translations. Translated by E. F. J. Payne. Total of 1,269pp. 5⅜ x 8½. 2-vol. set. Vol. 1: 21761-2 Vol. 2: 21762-0

MAGIC AND MYSTERY IN TIBET, Madame Alexandra David-Neel. Experiences among lamas, magicians, sages, sorcerers, Bonpa wizards. A true psychic discovery. 32 illustrations. 321pp. 5⅜ x 8½. (Available in U.S. only.) 22682-4

THE EGYPTIAN BOOK OF THE DEAD, E. A. Wallis Budge. Complete reproduction of Ani's papyrus, finest ever found. Full hieroglyphic text, interlinear transliteration, word-for-word translation, smooth translation. 533pp. 6½ x 9¼. 21866-X

MATHEMATICS FOR THE NONMATHEMATICIAN, Morris Kline. Detailed, college-level treatment of mathematics in cultural and historical context, with numerous exercises. Recommended Reading Lists. Tables. Numerous figures. 641pp. 5⅜ x 8½. 24823-2

PROBABILISTIC METHODS IN THE THEORY OF STRUCTURES, Isaac Elishakoff. Well-written introduction covers the elements of the theory of probability from two or more random variables, the reliability of such multivariable structures, the theory of random function, Monte Carlo methods of treating problems incapable of exact solution, and more. Examples. 502pp. 5⅜ x 8½. 40691-1

THE RIME OF THE ANCIENT MARINER, Gustave Doré, S. T. Coleridge. Doré's finest work; 34 plates capture moods, subtleties of poem. Flawless full-size reproductions printed on facing pages with authoritative text of poem. "Beautiful. Simply beautiful."–*Publisher's Weekly*. 77pp. 9¼ x 12. 22305-1

NORTH AMERICAN INDIAN DESIGNS FOR ARTISTS AND CRAFTSPEOPLE, Eva Wilson. Over 360 authentic copyright-free designs adapted from Navajo blankets, Hopi pottery, Sioux buffalo hides, more. Geometrics, symbolic figures, plant and animal motifs, etc. 128pp. 8⅜ x 11. (Not for sale in the United Kingdom.) 25341-4

SCULPTURE: Principles and Practice, Louis Slobodkin. Step-by-step approach to clay, plaster, metals, stone; classical and modern. 253 drawings, photos. 255pp. 8⅛ x 11. 22960-2

THE INFLUENCE OF SEA POWER UPON HISTORY, 1660–1783, A. T. Mahan. Influential classic of naval history and tactics still used as text in war colleges. First paperback edition. 4 maps. 24 battle plans. 640pp. 5⅜ x 8½. 25509-3

THE STORY OF THE TITANIC AS TOLD BY ITS SURVIVORS, Jack Winocour (ed.). What it was really like. Panic, despair, shocking inefficiency, and a little heroism. More thrilling than any fictional account. 26 illustrations. 320pp. 5⅜ x 8½. 20610-6

FAIRY AND FOLK TALES OF THE IRISH PEASANTRY, William Butler Yeats (ed.). Treasury of 64 tales from the twilight world of Celtic myth and legend: "The Soul Cages," "The Kildare Pooka," "King O'Toole and his Goose," many more. Introduction and Notes by W. B. Yeats. 352pp. 5⅜ x 8½. 26941-8

BUDDHIST MAHAYANA TEXTS, E. B. Cowell and others (eds.). Superb, accurate translations of basic documents in Mahayana Buddhism, highly important in history of religions. The Buddha-karita of Asvaghosha, Larger Sukhavativyuha, more. 448pp. 5⅜ x 8½. 25552-2

ONE TWO THREE . . . INFINITY: Facts and Speculations of Science, George Gamow. Great physicist's fascinating, readable overview of contemporary science: number theory, relativity, fourth dimension, entropy, genes, atomic structure, much more. 128 illustrations. Index. 352pp. 5⅜ x 8½. 25664-2

EXPERIMENTATION AND MEASUREMENT, W. J. Youden. Introductory manual explains laws of measurement in simple terms and offers tips for achieving accuracy and minimizing errors. Mathematics of measurement, use of instruments, experimenting with machines. 1994 edition. Foreword. Preface. Introduction. Epilogue. Selected Readings. Glossary. Index. Tables and figures. 128pp. 5⅜ x 8½. 40451-X

DALÍ ON MODERN ART: The Cuckolds of Antiquated Modern Art, Salvador Dalí. Influential painter skewers modern art and its practitioners. Outrageous evaluations of Picasso, Cézanne, Turner, more. 15 renderings of paintings discussed. 44 calligraphic decorations by Dalí. 96pp. 5⅜ x 8½. (Available in U.S. only.) 29220-7

ANTIQUE PLAYING CARDS: A Pictorial History, Henry René D'Allemagne. Over 900 elaborate, decorative images from rare playing cards (14th–20th centuries): Bacchus, death, dancing dogs, hunting scenes, royal coats of arms, players cheating, much more. 96pp. 9¼ x 12¼. 29265-7

MAKING FURNITURE MASTERPIECES: 30 Projects with Measured Drawings, Franklin H. Gottshall. Step-by-step instructions, illustrations for constructing handsome, useful pieces, among them a Sheraton desk, Chippendale chair, Spanish desk, Queen Anne table and a William and Mary dressing mirror. 224pp. 8⅛ x 11¼. 29338-6

THE FOSSIL BOOK: A Record of Prehistoric Life, Patricia V. Rich et al. Profusely illustrated definitive guide covers everything from single-celled organisms and dinosaurs to birds and mammals and the interplay between climate and man. Over 1,500 illustrations. 760pp. 7½ x 10⅛. 29371-8

Paperbound unless otherwise indicated. Available at your book dealer, online at **www.doverpublications.com**, or by writing to Dept. GI, Dover Publications, Inc., 31 East 2nd Street, Mineola, NY 11501. For current price information or for free catalogues (please indicate field of interest), write to Dover Publications or log on to **www.doverpublications.com** and see every Dover book in print. Dover publishes more than 500 books each year on science, elementary and advanced mathematics, biology, music, art, literary history, social sciences, and other areas.